MW00856222

HOOF BEATS

The publisher and the University of California Press Foundation gratefully acknowledge the generous support of the Constance and William Withey Endowment Fund in History and Music.

The publisher also gratefully acknowledges the generous support of the Publisher's Circle of the University of California Press Foundation, whose members are Ariel Aisiks / Institute for Studies on Latin American Art (ISLAA), Marcy Krinsk, Michelle and Bill Lerach, the Estate of Frances Kathryn Mitchell, Marjorie Randolph, and Peter Wiley.

HOOF BEATS

How Horses Shaped Human History

WILLIAM T. TAYLOR

With illustrations by Barbara Morrison

UNIVERSITY OF CALIFORNIA PRESS

University of California Press
Oakland, California

© 2024 by William T. Taylor

Library of Congress Cataloging-in-Publication Data

Names: Taylor, William, 1989– author. | Morrison, Barbara,
    1953– illustrator.
Title: Hoof beats : how horses shaped human history / William Taylor ;
    with illustrations by Barbara Morrison.
Description: Oakland, California : University of California Press, [2024] |
    Includes bibliographical references and index.
Identifiers: LCCN 2024004758 (print) | LCCN 2024004759 (ebook) |
    ISBN 9780520380677 (hardback) | ISBN 9780520380707 (ebook)
Subjects: LCSH: Horses—History. | Horses—Social aspects.
Classification: LCC GT5885 .T39 2024 (print) | LCC GT5885 (ebook) |
    DDC 388—dc23/eng/20240220
LC record available at https://lccn.loc.gov/2024004758
LC ebook record available at https://lccn.loc.gov/2024004759

Manufactured in the United States of America

33 32 31 30 29 28 27 26 25 24
10 9 8 7 6 5 4 3 2 1

This manuscript is dedicated to those who inspired this journey but did not live to see the printed book at journey's end—my dog, Logan; my grandparents, Bob and Berta and Ethel and Park; Cousin Dick, Uncle Will, and Uncle Flint; and my mentors, Sam High Crane, Susanne Bessac, and Bob Stahl.

CONTENTS

Plates follow page 120

ACKNOWLEDGMENTS

Putting together *Hoof Beats* has in many ways been a lifelong journey, and the list of people to whom I am indebted is too long to fit easily within these pages. Many wonderful people—especially my parents, Barb and Jim, and my early teachers, Elaine Kohler and Beth Kennedy—helped grow my curiosity and inspired me to choose this goofy and impractical career. Some, especially Terry Falcon, Aaron Rushing, Matt Stergios, and Roy Grow, taught me discipline and persistence. Others, especially Jorge Bravo, Bill Fitzhugh, Jim Dixon, and Bayaraa *akh* (Jamsranjav Bayarsaikhan), carried me into archaeology with their thoughtful mentorship. None of this work would have been possible without my beloved *bagsh* Tsermaa, my *profesora* Maria, and my ever-present guiding light through the dark recesses of academia, Emily Lena Jones. My siblings Matt, Cec, and Joc have been there since the beginning, and many friends—Brian, Connor, Gandhi, James, and everyone in the PHPR among them—have been there nearly as long.

The fellowship from DnD (Fernando, Jessie, Lauren, Joe, Alyssa, Logan, and the kids), and key friends like Isaac, Peter, Tunga, Julia, Avery, Alicia, Ryan, Ari, Arina, Svetlana, Oshan, Zahir, Björn, Mattheus, Rob, Jessie, Nik, and Shevan carried me through difficult times. Many senior scholars including Tina W., Pat, Jaelyn, Carole, Matt, Scott, Joe, Jack, Greg, Russ, Mindy, Bruce, and Gerardo offered kind guidance when it was badly needed, as did my little dingoes, Indiana and Ducky. My partners Jerry (Tuvshinjargal), Yvette, Ludovic, Brandi, Sarah, Chengrui, Yue, Rowan, Greg, Igor, Juan, Gustavo and Fito, Luna, Christina, all of the Nicks, Mark, Craig, Carlton, Gino, Pauline, Jessica, Sebastien, Antoine, Choongwon, Mike, Aida, Tina R., Kristine, Jimmy, Kristen, Tiina, Mijjee, Manabu, Angela, Victoria, Vera, Ankhsanaa, Bekbolat, Sukhee, Nicolas, Woody, Scott, Patrick, Joan, and countless other friends have shared their brilliance along with their blood, sweat, and tears as collaborators. My dear mentor, Nancy Stevens, helped me stay inspired. Wise words from friends, especially Emily, Nick, Rob, Juan, Christina, Jaelyn, Jimmy, Akin, Charlotte, Cyler, Victoria, Lee, and my beloved Mika, helped shape this manuscript into something legible. More dedicated archaeologists, scholars, and museum professionals than can easily be counted here toiled to help produce the work that this book is based on.

I am grateful to the expert artists who helped share this story, especially my amazing mother, Barb, along with the incredible Bill Nelson, Paula López Calle, Judy Peterson, Mark Williams, and Daria Chechushkova. Sabine Reinhold, Lars Holger Pilø, Espen Finstad, Pat Doak, Ganbaatar Galdan, Akin Ogundiran, Anatoliy Kantorovich, Jill Weber, Glen Schwartz, Manabu Uetsuki, Maebashi City Board of Education, and Johannes Eber graciously shared images, many from their own work, for this book.

The research for this book was funded with the gracious support of the National Geographic Society, the Fulbright Program, and the National Science Foundation (Award #1949305). I thank my amazing students, my wonderful colleagues at CU, and my dearest friends at the National Museum of Mongolia and the American Center for Mongolian Studies. I am especially thankful to those who have inspired me through their kindness and to those who build others up instead of tearing them down. To everyone whose contributions have escaped mention here, I am eternally grateful to you all.

PRELUDE

In the heart of the Mongolian steppe, in the cold interior of eastern Eurasia, sits the grassy Khangai mountain range, home to one of the world's last true horse cultures. Mongolia is changing fast; the country's bustling capital is a hodgepodge of business, education, international trade, and migrants who have left the countryside for urban life. In rural areas, motorcycles are seen in greater numbers each year as herders, scrambling to keep up with the demands of a globalizing and warming world, seek alternative modes of travel. But in the Khangai Mountains, life still moves to the beat of horse hooves. At the southern edge of this high range, there is a nondescript grassy knob overlooking a small bowl-shaped valley called Morin Mort, which translates roughly to "place of the hoofprints."

I have come to Morin Mort with a small multinational team of researchers in search of clues to the early history of people and horses. By early September, when our team arrives, the valley is nearly empty, with most of the local herders having already moved to their fall camps some distance away. The light drifting of early snow that dusts our camp one evening speaks to the logic and precise timing of their migration.

On this chilly fall morning, the valley, although now less populated, is far from empty. In fact, it is filled with clues to the earliest history of domestic horses. Around 1000 BCE, horsemen of Mongolia's first horse culture erected a series of monuments known as deer stones, large stelae sometimes three or four meters tall thought to depict ancient warriors. Each is equipped with necklace, earrings, and a tool belt (daggers, bows, shields, and even small horses) alongside elaborate deer images from which their name derives. Some scholars think that these deer images were meant to represent tattoos. Each stone is unique and probably represents a specific person, a brave warrior fallen in battle perhaps or a family ancestor (see plate 1).

It is not the deer or the stones, however, that have brought us to Morin Mort. It is the horses. Surrounding each monumental stone at Morin Mort is a small ring of individual stone mounds. Inside each mound is the head, neck, and hooves of a horse. During a ceremony, stones were erected, and each horse was killed, skinned, and eaten before being carefully buried around the stone, probably to guide or carry the deceased in the afterlife. Today at Morin Mort, only two stones remain standing. However, based on our survey and excavation of the site's remaining architecture, we can guess that when the site was constructed, there were at least five stones surrounded by more than one hundred individual horses. While this seems impressive, the number of such stones at other sites dating to this time is staggering, reaching into the hundreds and even thousands. Three millennia ago, horses were the beating heart of both economy and culture.

Atop a small ridge overlooking the stones and burials, we find hoofprints. At most times of the day, the small flat outcrop of blue-gray sedimentary stone is barely noticeable; a passerby is likely to walk over them without a second thought. But at dawn and again at dusk, something remarkable happens. With the sun on the horizon, the drab stone panels are illuminated, and the dancing shadows reveal perfect hoofprints carved into the stone. The effect is astonishing. The panels have been transformed into something like a muddy racetrack after a heavy rain. When I first laid my eyes on the carvings, I could have sworn that they were fossil tracks of ancient animals rather than rock carvings (also called petroglyphs). At this hallowed place, the relationship between humans and horses is literally inscribed into the fabric of the landscape. Through the scientific study of places like Morin Mort, my many colleagues and I have tried to understand the origins of horse domestication and the ways that it has shaped the world in which we live.

Many scholars, historians, researchers, and storytellers on many continents have taken on the task of exploring these questions. In my own retelling, I have drawn on a wide range of historical documents, linguistic analyses, oral traditions, even modern veterinary work. But our most important clues come from places like Morin Mort, places so ancient and so far from the Western world that they can be understood only through archaeology. It is precisely these pieces of the story that are the hardest to convey and the most commonly omitted from popular narratives and popular science. It is time to give archaeology its due, time now to use the scientific study of ancient horses themselves to tell the human-horse story.

Reading the archaeological record is a messy business, fraught with trial and error, mistakes, revisions, and debate. In the 21st century, the toolkits of the archaeologist have expanded from the trowel and brush into the realm of ancient genetics, stable isotopes, and biomolecules. These techniques are changing and diversifying at an astonishing pace. In this book, I bring together the latest knowledge from across the archaeological sciences to tackle old questions with fresh eyes and to rewrite pieces of the story that only a few years ago we simply took for granted.

This book is structured in four "hoof beats" corresponding to four key stages in the relationship between humans and equids. In beat 1, "Horses and People," I explore the origin and evolutionary history of horses and humanity's oldest relationship with the horse as predator and prey. In beat 2, "The Cart," I trace the earliest emergence of domestication. Based on a careful reanalysis of archaeological data and important new discoveries, I demonstrate how horse domestication began with the innovation of the chariot, spreading people and horses across much of Inner Asia. In beat 3, "The Rider," I explore innovation in horse control and how the emergence of mounted horseback riding transformed ancient Eurasia and repositioned the steppes as the center of global cultures, economies, and empires. Finally, in beat 4, "The World," I explore the journey of the horse over the globe's great oceans as they accompanied Viking explorers to the High Arctic, sparked drastic social change across the Americas, and revolutionized life in the colonies of Australasia. In the book's final chapter, I wrestle with the rapid disappearance of horses from a postindustrial world, and I outline how these "hoofprints" continue to shape our lives and our future.

In this journey through the archaeological record, I hope the reader will come away with both an appreciation for and a healthy skepticism of the contributions of archaeological science to our understanding of the world we live in. More importantly, I hope to stimulate readers to recognize the beats of the human-horse relationship that continue to reverberate in the world around us, from the Rocky Mountains to the Khangai Mountains and everywhere in between.

HORSES AND PEOPLE

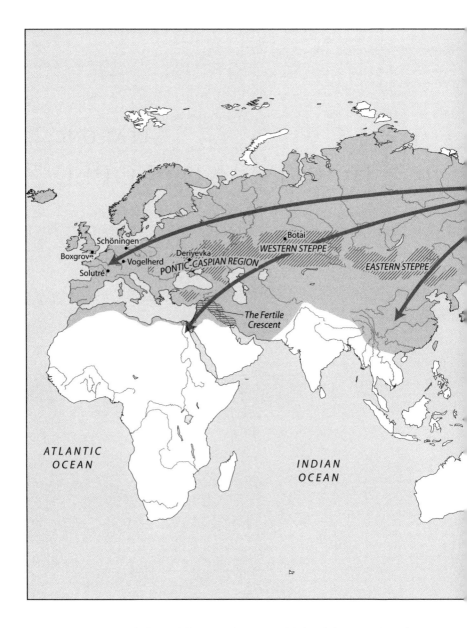

MAP 1. The evolution and dispersal of horses and their relatives out of North America. Map by Bill Nelson.

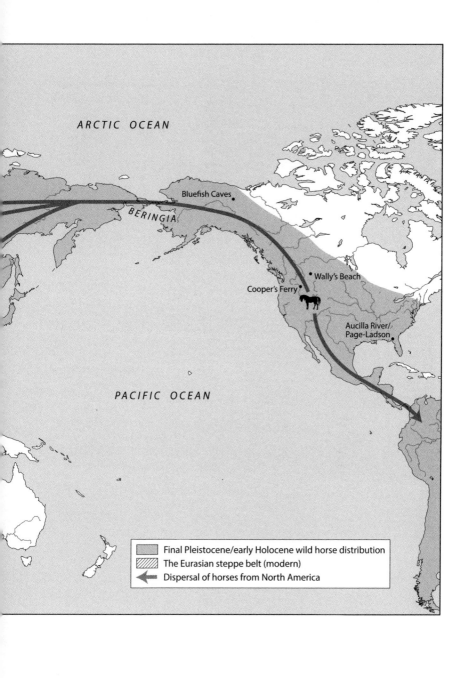

1 EVOLUTION

The story of the horse began with a bang—the biggest bang, perhaps, that our world has ever seen. Roughly sixty-six million years ago, a massive asteroid struck the Earth in what is now the Yucatan peninsula in the Gulf of Mexico. This cataclysmic event irrevocably changed the course of evolution and life on earth. In a geological instant, the toxic atmosphere, acidified oceans, and "impact winter" (a sudden temperature shift brought on by the obstruction of the sun's rays) that followed in the wake of the impact, many of the planet's animal and plant species were extinguished.

This collision is thought to have caused the Cretaceous-Paleogene (K-Pg) Extinction Event, a mass extinction that brought about the end of most of the dinosaurs. In the desperate ages that followed the asteroid impact, large creatures that roamed the desolated landscape either went extinct or were forced to adapt to a global ecosystem drastically shaken from its lowest ecological foundations.

In the chaos, new opportunities also opened. The startling scale of these extinctions left many ecological vacancies formerly filled by extinct animals, ready to be filled by those that had survived. In a process known as adaptive radiation, some groups of animals, including mammals and avian dinosaurs (birds), rapidly diversified to fill the ecological landscape vacated by those that perished. Fossil evidence shows that during this period, called the Paleogene radiation, a tiny, shrew-like arboreal creature—the first recognizable primate—flourished in North America and Europe, the progenitor of what would one day become humankind.

THE "DAWN HORSE"

After the Extinction Event, in a period known as the Eocene, a very different group of animals appears in the fossil record of the northern continents: the "dawn horse." Members of the order Perissodactyla, or odd-toed ungulates, these herbivores had three toes in the hind limbs and four in the front. Although scientists refer to them as "equids" in recognition of their role as the oldest members of the family Equidae, dawn horses were nothing like the tall, muscular beasts that we know today. The first dawn horses were probably browsers, meaning that they subsisted on leafy forage and brush from forest environments. They were also tiny; the first well-described dawn horse known from the fossil record, classified today as *Eohippus*, stood only around a foot tall.

In the post-impact world, both the dawn horse and proto-primates made their living in the forests and forest margins of a world recovering from catastrophic extinction. Although early perissodactyls were already quite different from early primates, Eocene horses had some surprising similarities with their tree-climbing brethren, particularly in the shape of their teeth. Finding the tooth of a relative of the dawn horse, early paleontologists once incorrectly took this discovery as evidence that monkeys once lived in British Isles.[1]

Over the millions of years that followed, the first ancestors of humans and horses would follow increasingly divergent evolutionary paths. Primates, adapted to flourish in forest environments, thrived in parts of Africa. Some of these species developed high intelligence and bipedal locomotion, perhaps adapting to a coevolutionary relationship with fruiting plants.[2] Within the last five million years, some bipedal primates left the closed forest and adapted to the savanna, becoming skilled scavengers, hunters, and gatherers. Sophisticated toolmaking skills, intelligence, and cooperative behavior made early humans the top predator nearly everywhere they went.

The evolutionary path of the dawn horse would also take it out of the forests, albeit much earlier than our own lineage. More than fifty million years ago, the tiny dawn horse was already excellent at a skill that would define many of its descendants—running. *Eohippus* was equipped with long, slender limbs capable of long strides and a spine design that kept its head stable while running.[3] The descendants of the dawn horse would come to include some of the world's fastest animals, fueled by the emergence of yet another type of new ecosystem that would redefine equid evolution and human history: the grassland.

THE EVOLUTION OF GRASSLANDS

If horses and humans are the lead characters in this story, then grasslands are undoubtedly the stage. Today, grasses (stalky, vascular plants belonging to the family Poaceae) include everything from the Kentucky bluegrass of suburban American lawns to bamboo, wheat, and the crops used to make the bread in our kitchen pantries. Members of this family live in arid, well-drained environments, and their high silica content makes them hardy and difficult to chew, digest, or destroy. Originally a tropical plant, grasses also benefited from the mass extinction event, becoming widespread, but not predominant, in the postapocalyptic landscape.

At first, grasses might have operated as minor parts of forest ecosystems, living in forest openings and understories.[4] Although the causes of this transition are debated, by around twenty million years ago, new ecosystems that were dominated by grass proliferated across the open habitats of the world, occupying vast stretches of land on some continents.

These green "seas" offered a wealth of plant food for the enterprising herbivore—if they could survive it. This was no easy task; for an organism to get enough energy to survive a diet of such harsh, high-silica materials requires huge input volumes. With the abrasion from grass, along with other dust and grit common in arid habitats, each bite also causes serious damage to the teeth, which are necessary to grind and chew rough grasses before they can be processed by the digestive system. Grasslands also offered major safety concerns: few accessible sources of water and exposure to extreme hot and cold temperatures. With no vegetative cover, all but the smallest herbivores would have been dangerously exposed to predators. To survive in the emerging prairies of the ancient world, herbivores needed innovative solutions (see plate 2).

A GRASS-POWERED MACHINE

Horses were once considered a textbook example of linear, directional evolution—a glorious sequence of improvements to the primitive *Eohippus* that culminated in the modern horse. However, we now know that this line of thinking is flawed. Evolution is messy and works through incremental adaptions to new circumstances, ambivalent to any long-term goals. The family tree of perissodactyls looked less like a linear progression and more like a braided stream filled with animals large and small that were adapted and adapting to an array of habitats and ecological niches.[5] From the Eocene, the perissodactyls became a diverse group that exploited a wide range of habitats, giving rise to animals such as the jungle-foraging tapir and the armored, wetland-loving rhinoceros. This diversification also produced the first proto-horses, strongly shaped by the special conditions of the open, dry grassland. Over millions of years, some early members of the horse family (see plate 2) developed behavioral and anatomic tools that enabled them to survive and thrive in the dry prairie. These adaptations fall into three general categories: energy, speed, and communication.

ENERGY

The most fundamental challenge of adapting to grassland was finding a way to eat it. How can a large organism live on a diet of thin, rigid plants so knife-like that they are referred to in English as "blades" and whose cellulose and silica walls make them exceedingly difficult to digest?

One answer lies in the horse's specialized teeth, which are the horse's first line of attack against grass's impressive defenses. Horses have pincer-like front incisors capable of snipping grasses from their stem, after which the plants are moved to their "cheek teeth," a long, single row of six molars and premolars on each side of the mouth. Each of these cheek teeth consists of a flat surface interwoven with sharp ridges of mineralized enamel. When horses chew using their characteristic circular chewing motion, the plant material is both crushed and ground, breaking down the tough protective coatings that make digestion difficult. After a few million years of grassland life, early horses also made a new and particularly critical innovation. Members of the proto-horse group *Equinae* became hypsodont, meaning that they developed high-crowned, or tall, teeth, which effectively served as hidden reserves of extra tooth above and below their gumlines. As grasses wore down the tough tooth surfaces over the course of an animal's life, ancient proto-horses were able to gradually replace lost tooth surface with reserves.

A second key to grassland energy among early horses was in their digestive system. Some large herbivores (among them cows, deer, and sheep) responded by innovating what is known as a ruminant stomach. The ruminant stomach uses a four-chamber structure that first ferments freshly consumed grass materials to break down their rigid components. Fermented material is regurgitated for more chewing before moving into lower chambers where nutrients are absorbed.

Horses have only a single-chambered, or monogastric, stomach, similar to that found in humans and dogs. On its own, the monogastric stomach is relatively inefficient at processing fibrous plant material, which speedily moves through the body undigested (this fact is one reason that fiber supplements are often prescribed for a blocked digestive system). How, then, are horses able to survive on a grassy diet?

Ancient perissodactyls and other mammals had a unique pouch-like addendum to the stomach known as a cecum, which also adapted to serve new digestive functions but without developing the multichambered ruminant gut. In horses, the cecum evolved to house a specialized microbiome capable of breaking down the tough bonds that make up cell walls in grasses and converting them into smaller, more easily usable compounds. In the horse's digestive system, dietary grass moves from the cecum into the colon, where after some last-minute resorption, it soon exits the body. The inefficiency of this system—relative to that of the ruminants—means that to meet their caloric needs, horses must graze continually, eating as much as 2.5 percent of their body weight in feed over the course of a day. Although teeth gave horses a powerful tool to deal with grassland environments, their need for continual grazing would influence their second major category of adaptations: speed and movement.

SPEED AND MOVEMENT

Even if grasses can be stomached, life in the prairie also requires water, and water sources are few and far between. Grazing enough territory to fulfill a horse's huge dietary volume quota thus requires endurance for long, regular movements. The exposed conditions of open grasslands also mean high risk of attack from predators, which can rarely be evaded through any other means besides speed. Early equids needed speed to outrun predators and the endurance to cover large swaths of territory while grazing.

Over the millennia, proto-horses responded to these challenges in myriad ways. Some were wildly successful; others resulted in extinction. Over time, ancestral horses converged on a body plan with longer, more robust limbs capable of immense strides, impressive musculature, and a specialized ligament structure that allowed them to stand continuously with little exertion but also to spring quickly into action. In the genus *Equus*, the three and four toes of the ancestral equids like *Eohippus* were streamlined down to a single hoof capable of sustaining the large, swinging strides of increasingly large-bodied animals.[6] This structure made them capable of fantastic stamina and velocity.

BEHAVIOR AND COMMUNICATION

While morphological and dietary adaptations were crucial to success in the grassland, behavioral adaptations were among the most important of all. Although speed is critical for evading predators, the protection offered by a social group can be even more impactful, especially for young animals trying to survive in an open prairie—the "strength in numbers" approach. Because of the advantages of group life in protecting the vulnerable, many medium-sized and large grassland herbivores evolved cohesive group structures and gregarious behavior, analogous to the social systems evolving in parallel among early primates and other mammals.

Present-day members of the genus *Equus* all display some form of group social structure, typically taking one of two forms. In the "territorial" system, found in donkeys and zebras (and more common in resource-poor and desert environments), females and their young aggregate into small clusters, with males competing among themselves to monopolize the largest territory encompassing the range of as many females as possible.[7]

In the harem system, found in domestic horses and their wild sister species (the Przewalski's horse), all the females and young animals live together, and a lone lead stallion lives with them. Unaffiliated males aggregate into bachelor groups ranging in size from a few to a dozen or more individuals.[8] Strange as this system may seem to people today, this structure is quite common among mammals, including primates: well-documented cases are found in langur monkey societies. Careful field studies show that living in these large social groups, wild horses form close friendships and that they coordinate their grazing and sleeping patterns closely, making them a cohesive and coherent unit capable of protecting each other from predators and surviving in the challenging environments of the steppe.[9] Horse social organization is also flexible, and some equids are even known to switch to a different social system when introduced to a new environment.[10] The evolution of a dynamic herd life helped ancestral horses survive in the exposed environments of the prairie and steppe, especially throughout the many fluctuations in global climate that have occurred throughout evolution during the Cenozoic.

Alongside these stable, long-term, and complex social relationships, horses also developed high intelligence and unique capabilities for communication. Perhaps because of their complicated social lives, modern horses have sophisticated social awareness, even eavesdropping on interactions between other horses within their group.[11] Horses are also capable of complex learning, memorization, and concept formation.[12] Although it is difficult to trace the antiquity of these social and behavioral systems in the fossil record, the key elements of horse society, including a hierarchy dominated by a lead stallion, sophisticated social relationships, excellent communication skills, and a tolerance for group life, likely helped the ancestors of horses navigate life in the grasslands of the ancient world.

HORSES, ASSES, AND ZEBRAS

While many lines of proto-horses came and went over the last fifty million years, genomic research suggests that all modern equids descend from a single line belonging to the genus *Equus* that evolved in North America around four million years ago.[13] Soon after, this lineage split into two deep divisions. These included the so-called caballine horses, that is, those that exhibit similarities to the modern domestic horse, on the one hand, and the zebras, asses, and donkeys, on the other.[14] The genetic differences across these two lineages today are significant, eclipsing the degree of separation seen between humans and chimpanzees (perhaps reflecting a wide range of adaptations to diverse environments).

Ancestral zebras and wild asses probably coexisted with early horses in North America exclusively until the Pleistocene, a mostly colder era characterized by numerous quick and dramatic fluctuations between ice ages and warm interglacials that lasted from roughly 2.5 million years ago until about fifteen thousand years before present. During the first of these ice ages, both the caballine and zebra/ass/donkey groups dispersed across the Bering land bridge into Asia, and by roughly two million years ago, into Africa.[15]

This Pleistocene era, with its sometimes severely cold, dry conditions, created difficulties for many animals and plants. Not so for equids, whose specialized adaptations to grassland life were perfectly positioned to benefit from these colder periods when temperate rainforests dried up and gave way to grasslands. In North America, stout-legged horses, including the Yukon horse, lived alongside the slender-legged animal with ass-like traits that some researchers classify within *Equus*, though others classify it in a separate genus—*Harringtonhippus*.[16] As animals spread out of North America, *Equus* had a mini-radiation of its own. Recent genomic sequencing suggests that Eurasian caballine horses split from their North American ancestors perhaps around one million years ago, although several later dispersals across the land bridge brought more recent intermixing episodes.[17]

In Asia, warmer southerly latitudes became the favored habitat of wild asses, which separated into a variety of subspecies, including the Tibetan *kiang,* the Mongolian *khulan,* and the Persian *onager,* while horse-like equids, including the now-extinct Siberian horse, Przewalski's horse, and the ancestor of the domestic horse, proliferated across the colder northern zones. In Europe, an as yet unnamed lineage of wild horses inhabited Iberia and perhaps other areas.[18] In Africa, asses and zebras adapted easily to the savanna. Africa developed two species of wild ass—the Nubian and the Somali—along with at least three species of zebra. And as an isthmus formed once again to connect the Americas across Panama, wild equids moved into South America as well, which boasted the small horse-like *Equus neogeus* and species of *Hippidion* (ancient relatives of *Equus*). There may well have been others, too, in all these places; population estimates using genomic data suggest that wild horses reached their population peak roughly forty thousand years ago, when the ancestor of the domestic horse split from the Przewalski's horse into separate species.[19] By the late Pleistocene, horses, asses, zebras, and their relatives lived on every continent except Australia and Antarctica.

SUMMARY

Over fifty million years, the descendants of the tiny dawn horse—a dog-sized creature browsing forest margins in the wake of the asteroid impact—had thus become consummate prairie animals, with specialized dietary adaptations for eating grass and elite anatomical and behavioral adaptations for speed and predator evasion in the open steppe. With the onset of the Pleistocene, the American *Equus* radiated across most of the world's continents, diversifying into an impressive range of asses, zebras, and horses. Occupying areas that could sustain larger groups, horses developed harem societies with rich systems of communication and cooperation.

In the colder conditions of the ice age, which kept the habitats of interior Eurasia cool and open, things were looking bright. As early members of *Equus* spread across new continents, though, they would encounter another child of the Extinction Event, descended from the tiny proto-primates that paleontologists once confused with dawn horses. This primate had now become a deadly, opportunistic carnivore with a taste for horse meat.

2 CONNECTION

As the world cooled, land bridges opened and early horses spread westward. As early as 2.6 million years ago, these hooved immigrants had reached Europe and the Indian subcontinent.[1] *Equus* soon diversified and adapted to their new environments, with early zebra-like species excelling in the African savanna and asses flourishing in the mid-latitudes of Africa and Eurasia. With their higher tolerance for cold conditions, horses thrived in the grass and forests of high-latitude Asia and Europe. Unlike the Americas, however, these new continents were full of hominins, ancestral humans like *Homo erectus* and their close relative. These early humans had spread out of Africa for the first time and were experiencing their own adaptive radiation across the world, from Africa in the south and Europe in the west, all the way to the coasts of East Asia. By somewhere around half a million years ago, the early hominin *Homo heidelbergensis* inhabited all the regions surrounding the Mediterranean. Expert toolmakers and proficient hunters of large animals, they became the first human ancestor to succeed in colder climates. Across Eurasia, the destinies of these early hominins and horses would forever intertwine.

It is difficult to say precisely when early humans first came into contact with early horses. Some evidence suggests that the very first humans to leave Africa might have interacted with relatives of the horse; in Georgia (the Eurasian republic between Russia and Iran), a large assemblage of fossil bones at the site of Dmanisi dating to around 1.7 million years ago included both *Homo* and *Equus* within the same deposit.[2] However, these Dmanisi equids were probably not true caballine horses, and there is no direct evidence that early *Homo* found at the site was hunting or eating them.[3]

Whenever the first relationship between humans and horses emerged, the archaeological record shows that horses may have been among the oldest animals to have been hunted by early humans. An early snapshot of the first interaction between humans and horses comes from the site of Schöningen, an open-pit coal mine in north-central Germany. Excavations revealed more than ten thousand animal bones along the shores of an ancient lake bed, along with stone tools and refuse associated with *Homo heidelbergensis*. The victims of this ancient slaughter were apparently ambushed while having a drink on the lakeshore, where options for escape were limited. Following the hunt, horses at Schöningen were subsequently butchered with the stone tools found at the site.

Another major archaeological discovery shows that Schöningen was not a one-time event and that horses played an even more complex set of roles in the diet and lifeways of early European hominins. Along the chalk cliffs of Boxgrove in the United Kingdom, archaeologists have uncovered the remains of a young adult female horse that had been hunted and extensively butchered. Just as at Schöningen, the horse appears to have been slaughtered using the tidal flats to pin the animal in an ambush, perhaps as she frequented a favorite freshwater drinking hole. Because the horse's bones were rapidly buried within the tidal flats, even the tiniest cut marks found on the surface of the bone were preserved.

By looking at the placement of these tiny marks, archaeologists can reconstruct the strategies and processes used by ancient hunters in butchering and using the horse after it was killed. Archaeologists determined that ancient hunters had not only removed the spinal cord, brain, and tongue of the Boxgrove horse but also cracked open most of the hollow bones to remove the marrow inside. Parts of the pelvis and other bones were also used to make "soft hammers," bone tools used for sharpening their stone knives as they dulled. The successful hunters, who may have been working together in a relatively large social group, then removed the rib cage and returned to camp with their meat. Perhaps most exciting among the finds at Boxgrove is a bone tool likely used to process the animal's hide. This artifact suggests that these early hominins may have worn horsehide clothing or produced other kinds of objects and garments from the animal's soft tissue.[4]

Since the site at Boxgrove is likely older than Schöningen by one hundred thousand years, these findings point to an even earlier point of contact between people and horses.[5] Taken together, these sites give us a close look at the use of horses by early humans and indicate that caballine horses were already a key part of the diet for human hunters by at least five hundred thousand years ago. Over the coming millennia, ancient horses would continue to be hunted by *H. heidelbergensis* and its Neanderthal descendants across a wide region stretching from the Altai Mountains of southern Siberia to western Europe.[6] When the first modern humans dispersed into northern Eurasia more than forty thousand years ago, they, too, would quickly develop a deep relationship with horses.

EARLY HORSE-HUNTING STRATEGIES

Because of their large body size and consistent behavior, among other valuable traits, horses were a reliable choice for human hunters moving into higher latitudes. The archaeological record suggests that anatomically modern humans hunted wild caballine horses in the horses' harem groups, using rivers, lakes, and natural features on the landscape to assist in driving and ambushing animals to a desired location. Paleolithic hunters often utilized natural gorges or benches, as well as water crossings, to ambush horses in groups, taking advantage of spatially restricted locations where the horses' incredible speed imparted little benefit without the room to run.[7] Because mass harvesting of horses often made use of fixed parts of the landscape, some hunting localities were probably used year-round, while others might have been exploited only during a particular season.[8] At the rock of Solutré in central France, horses were hunted in the same location across a stretch of roughly twenty thousand years or more, spanning the Middle and Upper Paleolithic (from between forty-three thousand and twenty-eight thousand years ago until approximately twelve thousand years ago).[9] At this spot, horses were diverted from the valley floor into a natural cul-de-sac and slaughtered en masse using the advantage of the natural topography.

At some archaeological sites of this type, archaeologists find the remains of entire bands of unlucky horses that fell victim to the hunt. In sites like this, almost all the parts of the horse are represented among the bones recovered, and very few animals besides horses will be found in the assemblage.[10] At Solutré, around 94 percent of all identifiable bones were horse bones. The demographic makeup of the horses found in a given site depends on which kind

FIGURE 1. Early hunters surprise a band of horses, trapping them against the edge of a waterway. Drawing by Barbara Morrison.

of group was hunted. When a harem group is slaughtered, the bones are likely to represent mostly adult females and juveniles, while bachelor-band hunting would produce all adult males, and sites reused over long periods over time for both types of group may develop a roughly even ratio of male and female horse bones.[11] These archaeological finds show that from the very beginning of their time at northern latitudes, people had an important relationship with horses as predator and prey.

EQUINE MASTERPIECES

Even as hunting of horses became an economic centerpiece of Paleolithic life in grassy areas of northern Eurasia, horses also entered into culture, artwork, and spirituality from a very early date. At the site of Vogelherd cave in southern Germany, dated to roughly thirty-one to thirty-two thousand years ago, archaeological excavations revealed more than seven thousand ancient animal bone fragments, of which horse and reindeer bones form the largest portion.[12] However, it is not the bone assemblage but rather the artwork that has made the site most remarkable. Among a series of remarkable objects found at Vogelherd are ten ivory figurines and pendants in the shape of Ice Age predator and prey animals, including mammoth, lion, bison, and horse. Perhaps the most famous figurine, known as the Vogelherd horse, is an intricately carved abstract horse decorated with tiny X incisions along its back representing a mane and tail.

Although the Vogelherd specimen is among the earliest examples of Paleolithic art, it represents only one instance of an emerging tradition of horse-focused artwork from across the continent. Horses were painted, carved, etched, and incised by Paleolithic hunters into nearly every available medium, including bone, stone, ivory, and presumably other soft organic materials that have not survived the test of time. In these representations, Paleolithic horses are shown running, feeding, mating, defecating, and being attacked by wild predators or hunted by humans.[13] Paleolithic horses are depicted in a variety of colors, from bay to black-dun, with zebra-like neck stripes and even leopard spots.[14] Many exhibit the mixed dun/white pattern still found on modern Przewalski's horses. The variety of horses in artwork faithfully reflected reality; through genomic analysis of ancient DNA preserved in Paleolithic horse bones to reconstruct the animals' original coat color, researchers determined that horses in the wild looked just as they were depicted in cave paintings.[15]

At the cave of Chauvet, in France, dated to roughly the same age as Vogelherd, beautifully realistic horses and other animals are painted along an undulating wall surface such that when lit by a flickering flame, they appear to move. Some cave paintings or artifact carvings incorporate such vivid anatomical detail as to depict specific facial muscle groups. In a database compiled of more than 4,700 images from European Paleolithic artwork, horses appeared most frequently, constituting roughly one-third of the total animal images and often taking center stage within a given panel (see plate 3).[16] Therefore, it appears that by the end of the last ice age, horses were already deeply embedded in human lifeways, economy, art, and culture in Eurasia. Horses were hunted using sophisticated tactics, sometimes in regular locations en masse, and they inspired a fantastic variety of carvings and artwork.

HORSES AND THE FIRST AMERICANS

Beginning around twenty thousand years ago, at the end of a time period referred to as the Last Glacial Maximum (LGM), the world began to warm. Rising temperatures and melting glaciers brought an end to the drier conditions that had helped *Equus* flourish across the grasslands spanning North America to Eurasia to Africa during the early Pleistocene. The vast steppes that supported ancient horse populations began to fragment, creating new opportunities and new challenges for both horses and people.

The changing conditions of a warming world also brought people and horses into contact for the first time in the Americas. During the LGM, North America remained connected to Eurasia between Alaska and Siberia through a bridge of elevated land covered in tundra-steppe vegetation. With this terrestrial link, the regions we now call Alaska and the Yukon are better understood as an ecological extension of the cold, high-latitude grasslands of Siberia than as part of North America as we know it today. This piece of "North American Siberia" was itself cut off, however, from the remainder of the Americas by large ice sheets that remained. Stone tools and butchered animal remains from the site of Bluefish Caves in Canada's northern Yukon now show that people lived in this area of North America by about twenty-four thousand years ago.[17]

Archaeological finds show that the first residents of arctic North America already had a relationship with the horse. Tucked in among the artifacts and bones found at Bluefish Caves is a broken segment of horse jaw, which shows distinctive cut marks recording the removal of the animal's tongue, a fleshy delicacy. As a warming climate melted the massive ice sheets that separated Beringia from lower latitudes, the first peoples moved southward, probably first making use of a coastal corridor that deglaciated earlier than an inland ice-free corridor that emerged later.[18] Once south of the ice, a relationship between people and horses persisted. Horse bones are found at some of the oldest archaeological sites south of the ice sheets, including Cooper's Ferry in Idaho, which can be dated to around sixteen thousand years ago, as well as at more than a dozen archaeological sites linked with Paleoindian hunters.[19] Horse bones are found at Paisley Cave, in Oregon, which has the ignominious distinction of being the home of the oldest human feces in the Americas.[20]

Wherever and whenever they were encountered, wild horses were important to the first peoples of ancient North and South America. At the site of Wally's Beach in Alberta, dated to approximately 13,300 years ago, Paleoindian hunters killed and butchered a group of seven horses, having trapped them near a waterway in a family band, just as had been practiced for millennia in Eurasia.[21] At other sites across Alberta, a well-dated set of caballine horse remains were found in direct association with other cultural material, including a cache of stone tools and several Paleoindian projectile points at the site of Brazeau Reservoir, dated to only 12,700 years ago.[22] Although the archaeozoological record selectively preserves a record of economic processes like hunting, horse skeletal remains also show clear evidence that horses had deeper and more complex social significance in the early Americas. For example, freshly fractured horse bones (that is, bone that was broken soon enough after the animal's death that the bone had not yet dried) were made into knives and tool handles by Paleoindian people at sites as far apart as Florida, New Mexico, and Nevada.[23] As migrants journeyed ever southward into South America, they also encountered both *Equus neogeus* and the donkey-like *Hippidion* (an ancient relative of *Equus*), whose remains are found at archaeological sites as far south as Argentina by thirteen thousand years ago.[24]

The first migrants south of the ice into the Americas found themselves in a dynamic, rapidly changing landscape. As the last ice age gave way to a warming planet, giant continental ice sheets ruptured, and the cold steppe-like conditions that dominated their margins across North America began to fragment and disappear too. Warming conditions and changing ecology wreaked havoc on the large herbivores of the steppe, including mammoth, camel, bison, horse, and the mega-carnivores that depended on them, such as saber-toothed cats and dire wolves. These environmental changes put escalating pressure on Ice Age herbivores, nearly all of which went extinct by the early centuries of the Holocene (the modern geological era, delineated at approximately twelve thousand years ago).

Ultimately, the stresses of rapid environmental change may have been too much for the American wild horse. In Alaska, horses appear to have declined consistently in body size, perhaps a result of dwindling access to food and habitat.[25] By ten thousand years ago, horses were no longer leaving a tangible presence in the fossil record of most parts of the Americas. Nonetheless, new research analyzing fragments of DNA found in sediments suggest that in some areas, such as the high latitudes of the Yukon, native horse species may have survived much longer, perhaps as long ago as five thousand years ago.[26] By the mid-Holocene, the currently available archaeological record suggests that most of the evolutionary homeland of *Equus* was probably horse free, although it is important to note that some scholars argue for a longer persistence—or even that horses were never lost in the Americas. In either case, the cultural memory of North America's ancient indigenous horses still persists today in many Native oral traditions.[27]

EXTINCTIONS IN EURASIA AND AFRICA

Back across the Bering Strait, in Eurasia, horse populations were also declining steeply as global climate change altered their habitats drastically. Analysis of genomic diversity in modern and ancient horses, which can be used to trace large-scale population dynamics in the past, suggests a precipitous population collapse in the proto-domestic horse, *Equus ferus*, beginning at around twenty thousand years ago.[28] During the Ice Age, practically all archaeological assemblages across Siberia contained horse bones, but climate warming and vegetation change turned grassland habitats into forests, and horses dwindled in frequency in human diets and in archaeological assemblages.[29]

Because the fossils of most Ice Age horses are difficult to tell apart just from looking at their bones, it can be difficult to understand exactly how many different horse species existed at the end of the Pleistocene. However, new results from ancient DNA suggests that since the LGM, many species may have gone extinct. Frozen remains of ancient horses from Siberia, preserved in permafrost or unique depositional environments, often preserve DNA that would otherwise have degraded. By studying well-preserved horse bones or even mummies, we now know that before the Holocene, at least one "ghost" species (probably *Equus lenensis* from paleontology literature) lived alongside *E. caballus* and *E. przewalskii*.[30]

As the periglacial steppes receded, so did many horses. *E. lenensis* moved northward, clinging to existence in the farthest northern edge of the Siberian Arctic until around five thousand years ago, when it seems to have finally gone extinct. Donkeys, wild asses, and zebras living in Asia and Africa had long ago adapted to desert conditions and thus fared much better in the warming Holocene, but at least one taxon of caballine horse occupying northern Africa probably also went extinct at this time.[31] Europe, too, lost a caballine horse relatively late in the Holocene, with at least one now-extinct species living into the 3rd and early 2nd millennia BCE in Iberia, and perhaps elsewhere.[32]

Despite the pressures of Holocene warming, Ice Age–adapted horses survived across much of Eurasia. In Europe from Iberia to the Baltic, wild horse taxa persisted in lower frequencies.[33] In Anatolia circa 6000 BCE, wild horses constituted nearly one-quarter of animal bones at some assemblages, declining to less than 2 percent between 4500 and 3000 BCE but persisting at least until around 2000 BCE.[34] In fact, some steppe areas retained a robust wild horse population well into the mid- and late Holocene, where horses accounted for as much as 40 percent of archaeological assemblages at various locations in northern and central Asia.[35]

SUMMARY

From the earliest use of high latitudes by ancient people, horses appear to have played a central role in subsistence strategies and other aspects of culture, from clothing to tools, artwork, and religion. Other early humans, like *Homo heidelbergensis*, engaged in sophisticated hunting strategies and used horses for a variety of other purposes, such as clothing, hundreds of thousands of years in the past. As Earth exited the LGM, environmental change threatened members of the genus *Equus* living at high latitudes, reducing their role in human diets around the globe. In North and South America, horses eventually disappeared from the fossil record, although new research suggests that they persisted much later in higher latitude areas. Other archaic species that once roamed the wilds of Europe, Siberia, and East Asia dwindled but persisted. Even amid declining populations in the warmer Holocene, wild horses remained a key part of many lifeways, particularly in the steppes of Inner Asia.

3 TRACING DOMESTICATION

Facing down a warming climate, declining populations, and disappearing habitat, the wild horse of the Holocene somehow emerged in a domestic partnership with people. But how? And why? Archaeologists and biologists have wrestled with these questions for decades. To answer them, we need first to grapple with the question of what domestication is and how it can be identified in the archaeological record.

At first glance, the difference between a wild or domestic animal seems intuitive. After all, many of us grew up having close friendships with a family dog or cat, which are then highly dependent on people in most aspects of their lives. We control what they eat (and where and when they eat it), we restrict their movements, and we choose whether they reproduce. We are responsible for our pet's health care. We may share a room, or even a bed. We might give our pet a name or dress it in a sweater when temperatures drop. We may even control when they die. At a glance, these deeply human-dependent lives are easily distinguished from that of a wolf or wildcat.

However, on closer examination, even the most domesticated animals actually enjoy a wide range of relationships with humans.

While a tiny Chihuahua might ride inside the purse of a Hollywood starlet, another member of the same species might be standing guard in the frigid Mongolian steppe, protecting against wolf attacks and rarely, if ever, actually entering a human household. In Australia, feral dingoes (descended from domestic dogs) live as wild carnivores, yet these same dingoes were cared for by Aboriginal peoples and trained to hunt game.[1] At any given time or place throughout the last few millennia, it is unlikely that any two dogs have had exactly the same relationship to humans.

The example of domestic dogs highlights the immense complexity and cultural variation that domestication truly entails, and many researchers now view domestication as a continuum rather than a binary state. Some domestic animals have several distinct genetic or behavioral differences from their wild progenitors; others have very few. Domestic animals that go feral, reentering the wild, can quickly experience drastic changes in their behavior or even appearance. While domestic house cats might enjoy cuddling, batting at Christmas ornaments, or occasionally even playing piano, feral cats are dangerous predators, killing so many small-bodied creatures in suburban and urban environments that their ecological impact can be catastrophic.

On the opposite side of the spectrum, even dangerous beasts like the elephant, grizzly bear, tiger, and orca can be captured, tamed, cuddled, or taught tricks; these so-called wild animals might live in a heated or air-conditioned house, be fed commercial food, or have regular veterinary appointments. Even some of the earth's most remote populations of wildlife, like wolves in Yellowstone National Park or penguins in Antarctica, have their movements, reproduction, and behavior monitored by radio collars or satellites.

Given this complicated variation across the animal kingdom, perhaps a more useful way to think of the study of animal domestication is not as a search for any particular trait or behavior but instead as a deep dive into the *relationship* between humans and animals and how it has changed over time.

THE FIRST ANIMAL DOMESTICATION

For most of human history, there were no domestic animals. Recent studies indicate that the first animal to develop a more intimate domestic relationship with humans was the wolf (*Canis lupus* or one of its close relatives), which appears to have first joined with human groups living in the colder climes of northeastern Eurasia.[2] As the great megafauna of the Ice Age dwindled across a warming continent, canines likely found safety and scavenging opportunities around human camps, where meat could be easily stockpiled in colder latitudes.[3] In return for their scraps and safety, dogs gave back to people; they minimized waste, helped protect the camp from predators and strangers, and became invaluable companions for activities like hunting and even transport. Although they might have entered the human sphere earlier, the first conclusively domestic dogs are found in the archaeological record of Eurasia by around sixteen thousand years ago and in the Americas by at least ten thousand years ago.[4]

Meat may have attracted the first dogs, but many of the animals most important to ancient humans were not meat eaters but herbivores. As a result, one of the most important steps in the domestication of animals was actually the domestication of plant crops. In the early Holocene, beginning around twelve thousand years ago, peoples living in the so-called Fertile Crescent and the adjoining Zagros Mountains (modern-day Iran, Iraq, Syria, and Turkey) began to cultivate the first domestic grains, such as wheat and barley.[5]

As cultivation of domestic plants became increasingly reliable, the people who cultivated them became tethered to particular spots on the landscape. With permanent villages emerging, local populations of game animals probably began to dwindle from over-hunting. Perhaps responding to the gradual depletion of wild game and bolstered by growing reserves of domestic grains, early farm-ers in western Asia captured and began to raise large western Asian mammal species, including sheep, cow, and goat, the first large domestic herbivore.

Genomic sequencing of some of the oldest domestic goats ever identified, from the site of Ganj Dareh in western Iran, suggests that their initial domestication probably involved only limited intervention, such as corralling and reproductive control through slaughter of young male animals.[6] These early livestock were prob-ably raised only for meat; only later did herders begin to breed them for other specific products such as milk and wool.[7]

This new economic system of domestic plants and animals had some significant cultural and social consequences. When disaster struck, such as extreme weather or disease, those with small herds or poor cropland suffered disproportionately. Herds and prime farmland or pasture was passed on to the next generation rather than shared, and the sons and daughters of unlucky farmers found themselves facing limited resources. As a result, domestication in the Fertile Crescent eventually prompted many to quite literally seek greener pastures through out-migration. Farmers and herders migrated from the eastern Mediterranean and Mesopotamia, bringing food-producing strategies and domestic animals into

FIGURE 2. Early pastoral economies developed in the Fertile Crescent and the Levant thousands of years before the domestication of the horse. Drawing by Barbara Morrison.

large swaths of Europe, Africa, and Asia. In other cases, economic strategies and plants and animals were voluntarily adopted by neighboring peoples without necessarily involving population movements, and by around 6000 BCE, domestic sheep had spread as far as Kyrgyzstan and Central Asia.[8] In China, independent domestication processes were also taking place, bringing grains like rice and millet into the human sphere, followed by the capture and raising of pigs.[9] The dramatic expansion of the agricultural and pastoral world brought domestic animals and plants to the edges of the Eurasian steppe belt, home to those wild horses that had adapted successfully to the warmer post–Ice Age world. It is here, at the margins of the changing agropastoral sphere, that archaeologists seek clues to the first domestication of the horse.

THE DOMESTICATION RELATIONSHIP BETWEEN HUMANS AND HORSES

Before we can identify horse domestication in the archaeological record, however, it is important to characterize the tremendous variety of roles and relationships that might fall under the umbrella of horse domestication. Today, the most iconic role of horses in relationship to people is their use in transportation, though they are also regularly used for trade, crowd control, and racing. They move agricultural equipment such as plows and threshing wheels and pull conveyances such as carriages and carts.

Domestic horses are also extensively managed in their day-to-day lives. A domestic horse's reproduction is often carefully controlled. Animals may be castrated or tricked into mating with an animal they would not otherwise pair with. People exert varying levels of control over a horse's environment and diet. A domestic horse may be fenced in a single location year-round, or they may graze semi-freely, only to be herded intentionally to specific pastures in specific seasons. A horse might live in a stall by itself eating an artificial diet of processed grain. People may groom and modify a domestic horse's body, trimming or braiding a mane or a tail or branding the animal to signify ownership. To mitigate health problems, domestic horses are often well cared for; human intervention may help heal a broken leg, extract a problematic tooth, or perform a medical procedure or medicinal ritual to cure a domestic horse of sickness. In some cases, horses are milked to produce the fermented *airag* or *koumiss*, or they may be slaughtered for their meat.

Characterizing the full range of horse usage across societies worldwide today facilitates the study of the biological and osteological consequences of domestication. The many roles played by modern horses provide us with a whole suite of possible care- and use-related processes to look for in the archaeological record, from raising to feeding, movement to milking. But without a time machine to observe these actual interactions, how can we confidently trace them in the deep, deep past?

TRACING DOMESTICATION IN THE EQUINE SKELETON

Over the years, archaeologists have explored many different proxies for tracing the earliest domestication, from ancient artwork to food crusts preserved on pottery sherds. But our best and most direct chance at reconstructing the early history of horse domestication comes from an imperfect and underutilized scientific resource: the bones of ancient horses themselves. Study of animal skeletal remains, a science known as archaeozoology, provides the most direct surviving insights into the ancient interactions between horses and people.

As the discipline of archaeozoology was gaining traction as a source of insights into horse domestication during the 20th century, few direct links were initially available to connect human activity with horse bones. As an initial obstacle, even basic identification of equid bones can be logistically challenging, since few reliable traits distinguish *Equus caballus* from their close relatives, like *E. przewalskii*. In cases when bones are not well preserved, the problem is even worse; horses are not easily distinguished from other equids, such as asses, donkeys, and zebras. Researchers can, of course, easily identify instances when horses have been killed and eaten by examining patterns of breakage, cutting, and butchery. However, patterns of butchery or food preparation are not much help in separating horse herding from the millennia-long tradition of horse hunting that characterized much of early Eurasia.

Lacking alternatives, most horse researchers looked for indirect lines of evidence for human influence, especially patterning in when, where, and how many horse bones are found at archaeological sites. Such information is interesting and important but very hard to link to human activity. For example, one common idea, which still persists today, was that a sudden increase in the frequency of horses or their appearance in new geographic areas might help trace early domestic horse assemblages. However, other nonhuman processes can easily influence the distribution or frequency of horse bones. For example, an increase in the number of horse bones at archaeological sites during a given period could reflect a greater number of wild animals on the landscape as a result of environmental changes or a growing role for wild horses in the diet of ancient hunters.

Another factor explored by archaeozoologists interested in domestication was bone size and shape. Some influential and creative scholars suggested that human control over reproduction and environment could have introduced changes in equine skeletal morphology.[10] Might an increase in the shape and size variability of archaeological horse bones help differentiate early domestic assemblages from the bones of hunted horses?

Unfortunately, on further scrutiny, each of these patterns also has other equally plausible explanations. Changes in the size, shape, and distribution of horse bones observed in early Eurasian assemblages may be caused by climate and environmental change. Across much of Eurasia, the climate has varied considerably during the last ten thousand years. Following the end of the last Ice Age, regional changes in vegetation, temperature, and precipitation had a big impact on the distribution of available habitat for horses.[11] After a peak of solar activity and a warm spike roughly nine thousand years ago, a broad cooling trend impacted much of Europe and northern Asia over the mid-Holocene.[12] The body size of wild horses is known to adapt to environmental parameters, typically decreasing during times of poor resource availability and expanding under favorable conditions.[13] As conditions changed and vegetation fluctuated between forest and grassland, increases in the frequency of wild horses on the landscape, changes in their body size, and the extent of their distribution could have easily produced the patterns once hypothesized to be evidence of domestication.

Anthropogenic environmental change caused by early (non-horse) pastoralists may also have impacted wild horse populations in the mid-Holocene. Beginning around 6500 BCE, agricultural economies spread out of the Near East and into Europe. As they went, they often slashed and burned forest, converting it into arable lands. This agricultural deforestation alone was probably sufficient to explain a substantial increase in the proportion of horse bones in assemblages.[14] In new grassland environments created by human forest removal, some wild horses would have received better nutrition and grown to larger sizes, while others occupying more marginal zones remained smaller. Consequently, changing habitat could have generated size variability among wild equids and increased the abundance of these animals across the landscape of ancient Eurasia.

Indeed, most mid-Holocene horse assemblages supposedly linked to early domestication also look suspiciously similar to hunted horse assemblages of earlier periods. Horse bones in early

Black Sea steppe sites are typically found with large numbers of other wild animals, but these are not found in association with domestic animal bones.[15] Before the 4th millennium BCE, wild horse bone frequencies in the Ural region of Russia ranged widely from a few percentage points to more than half of the total identified specimens, showing that opportunistic slaughter sometimes produced very high ratios of wild horse bones, even during the Holocene.[16] The regular presence of entire horse carcasses and a high frequency of adult animals found at late Neolithic sites are also classic traits of Paleolithic game-drive assemblages. While efforts to link changes in horse bone frequency and morphology to domestication still persist in archaeological literature, any actual link to a changing relationship with people is hard to establish.

EARLY MODELS FOR THE DOMESTICATION OF THE HORSE

Despite the difficulty of extracting reliable information from horse bones, researchers began to stitch together a plausible narrative for the early domestication of the horse, relying heavily on other kinds of indirect archaeological clues. Early on, one of the most influential sets of clues came from burials of the Yamnaya culture, distributed across a stretch of grassland north of the Caspian Sea spanning eastern Europe and western Asia known as the Black Sea steppe or Pontic-Caspian steppe.[17] During the 4th millennium BCE, burials of this culture show evidence of disassembled wheeled carts and domestic animals. During this era, Yamnaya and other archaeological cultures spread into new areas of eastern Europe and western Asia, with archaeological data showing that they raised domestic sheep, goat, cattle, and pig and cultivated domestic grains such as wheat and barley. Yamnaya people migrated impressive distances from their homeland in the Black Sea steppe, reaching eastward into the mountains of central Mongolia by around 3000 BCE—a distance of roughly half the continent—and westward into Germany and central Europe by at least 2500 BCE.[18] The migration of the Yamnaya people is also plausibly associated with the spread of Indo-European languages and new steppe-linked burial practices, including large burial tumuli known as *kurgans*.[19] For many years, researchers have hypothesized that this impressive set of social transformations was linked to the first horse domestication.

Some interesting changes also occurred in the archaeozoological record during the 4th millennium BCE. Following many millennia of post–Ice Age decline, many areas of central Europe, including southern Germany and the Balkans, saw an apparent spike in the frequency of horse remains found at archaeological sites, coupled with an apparent increase in their morphological variability.[20] In the absence of more direct indicators to rely on, this unexpected reversal of a steady post-Pleistocene decline was interpreted as evidence of early horse domestication.[21]

Other indirect archaeological clues were also added to the dossier supporting the Yamnaya domestication hypothesis. For example, horses are sometimes found in artistic imagery from the region, and equid bones were occasionally recovered within human graves, both of which imply a significant role for the horse in local cultures and cosmology in the Black Sea steppe.[22] Some objects from this period were hypothesized to be early bridle equipment, including perforated pieces of antler resembling the cheekpieces of a bridle.[23] Together, these indirect lines of horse evidence bolstered the case that the spread of the Yamnaya, Indo-European languages, and domestic animals across the continent during the 4th millennium BCE was stimulated by the domestication of the horse. Over the years, this model's simplicity and tremendous geographic scope has given it staying power, and it remains common in some corners of archaeological literature. However, establishing the validity of the Yamnaya model required direct evidence of human control of horses.

DIRECT MARKERS FOR DOMESTICATION

In recent decades, the maturation of scientific archaeozoology has provided us with a diverse toolkit for tracing human-horse relations in the deep past, especially through using diagnostic changes to the equine skeleton caused by human activity. Bones are strong and resilient enough to survive in archaeological deposits, and as remnants of living organisms, they directly reflect the life history of the animals that they once belonged to. With the right background knowledge and careful observation, the skeleton can provide a stunningly detailed snapshot of the activities, behaviors, and processes that played out over an animal's life.

Simple demographic patterning is an important and powerful tool to trace processes of domestic animal management. Raising a reproductive herd of horses usually requires killing or castrating young male animals, which disrupt herd hierarchy as they reach sexual maturity around the age of three years. Through careful study of ethnographic horse herders in Mongolia and Kazakhstan, researcher Marsha Levine identified patterning in the age and sex of animals from a pastoral herd.[24] Adult females are kept for breeding, and except for one or two stallions per herd, adult males are usually castrated. Except for when they are slaughtered for special occasions, such as when someone is buried with their horses, pastorally managed horse assemblages thus usually contain very few breeding-age adult animals, especially prime-age breeding females. In archaeological deposits of domestic horses, we instead find an unusually high proportion of elderly animals, mostly female, and of juveniles that have not yet reached breeding age, mostly male.[25]

These demographic patterns present a strong contrast with hunted horse assemblages, which tend to yield a high fraction of healthy adult animals. If a site formed through the hunting of family groups/harems, bachelor bands, or both, these hunted assemblages might be mostly female animals, mostly male animals, or an even number of both. Crucially, though, pastoral assemblages will almost never include large numbers of breeding-age adult female animals, even when ritual activity is involved. Therefore, by reconstructing the age and sex patterns of archaeological bone assemblages, researchers can assess whether an assemblage was likely generated through hunting, herding, or other processes.

The quest for answers to early horse domestication was revolutionized by the discovery of other direct indicators of human control. Initially pioneered by scholars such as Juliet Clutton-Brock and Mary Littauer, who worked on animals from ancient Egypt, archaeologists recognized that horse transport could sometimes cause unique damage to a horse's teeth, a phenomenon known as bit wear.[26] This type of damage occurs when a bridle mouthpiece, known as a bit, contacts the chewing surface of the lower premolars of the horse during riding. If the bit collides with the teeth during transport or the horse attempts to fight or escape the effects of the bit by taking the bit between its teeth, the harder metal surface can wear the softer tooth surface in a characteristic way. Traumatic interactions like this are rare in many contemporary riding systems, which depend more on training and nonphysical cues. However, bit damage still occurs today in bridle systems that rely heavily on mechanical forces for horse-rider communication and control.[27] Building on the bit-wear research of Clutton-Brock and others like David Anthony and Dorcas Brown, archaeozoologists have now developed a wide range of scientific techniques directly tracing human activity in the horse skeleton.[28] These tools allow us to directly identify horse transport in archaeological assemblages. Through careful comparison of the skeletons of wild and domestic horses, we have learned that metal mouthpieces from a bridle can not only damage the teeth of horses but can also crack, chip, and deform the bones of the jaw.[29] A horse that has been bridled or haltered often develops faint but clearly recognizable grooves in the bones of its nose that can be seen with a 3D scanner.[30] Cracking, fusion, and bony growths in the mid-back are found in horses that have been ridden, especially those with only a soft pad saddle (see plate 4), while those that have been used to pull chariots or carts develop problems in the neck and shoulders and may sometimes develop unique damage to the upper dentition.[31] Horses that are ridden with a bridle that leverages the jaw may develop osteoarthritis, and horses that are heavily exerted develop specific osteological changes to their premaxilla, where the rigid muscles involved in breathing come into contact with the bones of the nose.[32]

The processes of horse care and control over movement, too, are often recorded in bone. Horses develop pathologies specific to living in confined spaces.[33] When sick or wounded, domestic horses are often cared for by humans in ways that can be observed in the skeleton, through healing or surgery.[34] Measurements of isotope ratios in carbon, oxygen, nitrogen, and strontium molecules mineralized inside the enamel of a horse's teeth can be used to identify when a horse was fed grain in the winter, consumed local water sources, or was traded to a new region in its childhood.[35] When preservation is good, ambiguous distinctions between wild and domestic horses based on morphology can now be made conclusively using ancient DNA.[36] Clues to ancient herd-management practices, including genetic selection for specific traits like coat color or gait, are encoded in DNA and can be reconstructed in a laboratory. This rich scientific toolkit now provides researchers with the ability to clearly and directly identify domestic horse transport, reproductive management, and care in archaeological assemblages.

DIRECT SUPPORT FOR THE INDO-EUROPEAN HYPOTHESIS?

Bit wear provided one of the first direct avenues for tracing horse transport in the archaeological record. When David Anthony and his colleagues identified a case of bit damage on the skull of a male horse from the site of Deriyevka, Ukraine, it appeared to provide hard evidence for the early domestication of horses in the Black Sea steppe.[37] Based on stylistic comparison of artifacts found at the site, this find was thought to date to the Neolithic/Bronze Age transition, around 4000 BCE. If true, the Deriyevka horse not only would support the idea that early Yamnaya herders domesticated horses and used them to disperse across Eurasia, but also suggests that mounted riding might predate even the invention of the wheel itself, which first appeared in the archaeological record centuries later.

Sadly, this story was too good to be true. Direct radiocarbon dating later revealed that the Deriyevka horse dated to between around 700 and 200 BCE, making this find little more than an intrusive, Iron Age burial of a domestic horse deposited into a pre-existing and more ancient site.[38] With Deriyevka relegated, the Yamnaya hypothesis was left where it began: a compelling possibility but still lacking any conclusive proof.

ALTERNATIVE EXPLANATIONS

If Deriyevka, Yamnaya, or other early horses from the Black Sea region were not ridden (and perhaps not even domesticated), what would explain the various lines of evidence that seemingly pointed to domestication, such as large-scale changes in frequency and morphology of horse bones, horse-related artwork, and apparent horse equipment?

As shown in chapter 2, the presence of horses in artwork or their inclusion in ritual deposits also fails to differentiate late Neolithic Black Sea assemblages from earlier Eurasian societies that hunted horses.[39] Indeed, paintings, carvings, and ritual depictions of horses have been perhaps the most deeply rooted tradition in hunter-gatherer artwork across much of northern Eurasia over the last thirty thousand years.

Finally, more careful scrutiny of apparent riding equipment shows that such objects are poorly linked to horses. Perforated antler tines once thought to be used as bridle components are rarely if ever found in clear association with horses.[40] Usually boasting a single hole, these differ fundamentally in their design from later multi-hole cheekpieces, or psalia, used as components of horse bridles and are just as likely to be parts of composite tools.[41] Similarly, looped copper rings found in burials of the Caucasus Maikop culture to the south of the Black Sea steppe and sometimes described as bridle cheekpieces have recently been found in situ to be nose rings used to guide domestic cattle.[42]

A SECOND ENEOLITHIC CANDIDATE: BOTAI

With the erosion of Deriyevka as a candidate for early horse domestication, scholarly attention in recent decades has shifted away from Ukraine eastward to the northern margins of Kazakhstan, to an archaeological site known as Botai. At this ancient village, dated to roughly 3500–3000 BCE, researchers recovered a large faunal assemblage consisting of more than 99 percent horse remains. Like Deriyevka and the Yamnaya cultures, Botai was situated at the edges of the expanding pastoral world of eastern Europe and western Asia, in prime habitat for wild horses.

Upon its initial discovery, many of the same considerations that led researchers to argue for a Yamnaya horse model also seemed to provide compelling evidence for domestic horse use at Botai. Excavations at the site yielded some intricately carved horse bones and some apparent ritual features (such as one human burial surrounded by horse remains).[43] At Botai, horse bones were used to manufacture tools, and excavations revealed post holes, hypothesized as a possible corral, as well as large deposits of decayed organic material within a house structure.[44] Although none of the horse teeth found at Botai showed classic damage to the upper surface of the premolars from metal bits, Anthony and Brown hypothesized that organic bridle mouthpieces might cause a more equivocal form of damage on some Botai horses.[45] Once again, circumstantial evidence appeared to be gathering in support for horse domestication in the 4th millennium BCE. However, once again, direct indicators of domestication remained elusive.

DIRECT EVIDENCE FOR DOMESTICATION AT BOTAI?

Using these methods, both old and new, modern archaeologists are now armed with a range of techniques to find the origins of horse domestication using direct lines of evidence. However, these lines of evidence have produced very little to support the argument for early domestication in the Yamnaya or Botai cultures.

Analyzing the vertebrae assemblage from Botai, Levine found no evidence of vertebral damage caused by horseback riding, other than a few small changes consistent with natural aging.[46] The apparent bit wear identified on Botai horses by Brown and Anthony was soon debunked when archaeozoologist Sandra Olsen demonstrated that natural tooth wear in wild horses could produce essentially identical patterns, making these features unreliable as a marker for horse domestication unless analyzed in tandem with the opposing tooth.[47]

From a horse rearing and care perspective, too, the Botai assemblage produced results that were wildly inconsistent with domestication. Researchers found that Botai horses could not be differentiated in terms of size or shape from wild horses found in earlier periods and showed no evidence for morphological change caused by domestication.[48] As shown by three independent studies (using both skeletal morphology and DNA), the Botai horse assemblage is inconsistent with the expected demographics for a managed herd, exhibiting a roughly equal balance between male and female horses, with the age of most horses falling between three and eight years.[49] This population structure does not match any pastoral assemblage from later periods that had been managed for

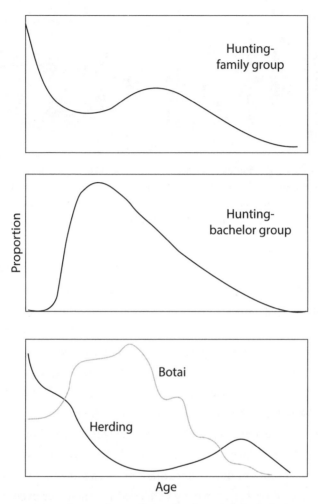

FIGURE 3. Expected mortality profiles for archaeological assemblages arising from hunting of family groups (*top*) and bachelor bands (*center*), along with expectations for a managed pastoral herd being culled for meat (*bottom, black*) as compared to Botai (*bottom, gray*). Image by author.

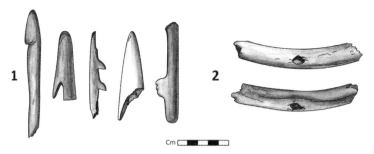

FIGURE 4. Bone/antler projectiles found in excavations at Botai, showing impact damage (1), along with horse rib bones showing evidence of impact during hunting (2), modified from S.L. Olsen's "The Exploitation of Horses at Botai, Kazakhstan." Image by Mark Williams.

either meat or milk, or both, since the slaughter of fertile adult animals has a catastrophic effect on herd reproductive dynamics.[50] Instead, this pattern closely matches that of many other assemblages of hunted animals from Paleolithic Europe, particularly those where a site was reused over time for the hunting of both harem (mostly female) and bachelor (mostly male) groups.

Finally, the Botai assemblage includes indisputable proof that the horses were hunted, with several projectiles found in association with the horse assemblage, and that at least one of the horses had actually recovered with a bone dart lodged in its ribcage. Based on the absence of these direct markers, even proponents of Botai domestication have typically conceded that many or even most horses at Botai were hunted.[51]

TOWARD A BOTAI CONSENSUS

Despite these significant gaps in the argument for domestication at Botai, scholarly opinion swung drastically in favor of this hypothesis in the first decade of the 21st century, with the publication of a high-profile paper that seemed to demonstrate conclusive evidence of bit-wear damage to a Botai horse tooth and ceramic residues pointing to the use of Botai horses for milk.[52]

To begin with, scholars analyzing horse remains from Botai using the shape of the metapodial (a lower limb bone) sought to demonstrate that the horses were not wild animals but were instead the domestic *E. caballus*. Measurements were taken and compared statistically with Ice Age wild horses, modern Mongolian horses, and Bronze Age domestic horses, and these results seemed to show that the Botai horses were closest to domestic comparatives.

While earlier arguments for bit wear at Botai had proven to be unreliable, researchers identified new, more reliable forms of bit wear occurring on the front margin of the horse's second premolar that could be more confidently linked to a metal bit.[53] Reanalysis of the horse specimens from Botai revealed a single tooth apparently exhibiting this form of wear (parallel-sided in shape, with visible exposures of dentine, the softer material underlying the tooth's hard enamel). Several Botai horses also showed small, raised plaques of bone along the diastema, or bars, of the jaw that seemed to be more developed than those usually found in wild animals.[54] Together, these two more convincing forms of bit wear helped sway scientific opinion that some of the Botai horses were actually ridden.

Another key assertion from this study was the suggestion that residents of Botai regularly consumed horse milk. Looking at ceramic potsherds from the site, the team first compared the isotope ratios of residues found on these ceramics with known reference values (from sheep, goat, cow, pig, horse, and fish fat), showing that organic residues found on Botai pottery were likely left over from ancient horse fats (a fact that was not surprising, given that the site's animal composition was 99 percent horse remains).[55] In a later update, researchers sought to distinguish whether these residues were caused by meat fats or milk fats. Although the carbon isotope values they measured could not separate the two, they found a strong signature of summer seasonality in a hydrogen isotope known as deuterium. Based on the apparent summer seasonality signature, they argued that the residues were from milk, which is produced in the summer months. However, to date, no direct traces of milk, such as ancient milk proteins, have ever been recovered from these ceramic fragments.

Armed with this new evidence for riding and milking, the Botai findings tipped the scales toward widespread acceptance of these horses as the earliest evidence for horse domestication. Longstanding criticisms of the assemblage, such as the absence of vertebral pathologies and the lack of demographic evidence for culling/reproductive control, were mostly forgotten. The Botai-first domestication model became a new, modified version of the Yamnaya hypothesis. In the Botai model, early domestic horses from Botai were transferred to the Yamnaya and other Indo-European speakers, who then used them to migrate across northern Eurasia during the 4th millennium BCE. In many ways, this argument remains the predominant explanatory model for horse domestication in use today.

EMERGING TECHNOLOGIES UPEND THE BOTAI AND YAMNAYA NARRATIVES

Since the publication of the initial Botai findings, however, the acceleration of research technologies in archaeological science has produced a wide range of insights that wholly contradict the conclusion of domestication at Botai. As tools capable of sequencing and capturing ancient DNA have proceeded by leaps and bounds, complete sequences of the nuclear genome can now be generated from isolated pieces of horse bone. Whereas early scholars struggled to distinguish caballine horses on the basis of their morphology, ancient DNA now enables precise identification of species, detailed comparison of ancestral relationships, and identification of physical traits using only fragments of bone and tooth.

Despite the early argument to the contrary based on metapodial shape, a startling 2018 finding and full-genome DNA analysis of the Botai horses revealed them to be *Equus przewalskii*, the closest relative of the domestic *E. caballus* but one that has never been managed as a domesticated animal.[56] This wild horse is aggressive, intolerant of human presence, and extremely poorly suited to riding, and it had never in recorded history been used for transport.[57] Later the same year, a second study evaluating human DNA from Botai individuals showed no connections between Botai and Yamnaya people, the supposed recipients of their innovations with horse riding.[58] In just two strokes, key assumptions behind the Botai and Yamnaya hypotheses (namely, that Botai animals were the progenitors of modern domestic horses and that they prompted the widespread social transformations of the 4th millennium BCE) had been completely falsified.

With an increasingly detailed understanding of horse dentition, the apparent evidence for transport at Botai has also withered. As an archaeozoologist specializing in the study of horse domestication, my own research has focused heavily on improving our methods for tracing horse transport in the skeleton of the horse and separating natural from cultural processes. In 2021, I published a collaborative paper with Christina Barrón-Ortiz of the Royal Alberta Museum, a leading specialist in the osteology of wild horses who has spent years analyzing the development and wear of teeth of ancient horses from Pleistocene North America. Our comparison of the Botai horses with these wild equids showed that the damage identified on a single tooth at Botai that had been ascribed to bit wear was more likely caused by natural phenomena and that similar features were visible on wild horses from Pleistocene North America.[59] Specifically, small divots in the enamel interpreted as bit damage may actually be pit-form enamel hypoplasia, a kind of dental defect caused by a disruption in enamel formation during tooth development. We also showed that even in wild horses, exposure of the enamel on the front edge of the premolar can be caused by natural wear in areas of thin cementum, which otherwise covers the tooth crown. Finally, we found that small amounts of mandible bone formation, of the same severity of those linked to bit use in Botai, are also found in wild horses hunted by early North American groups. With these new considerations, the Botai assemblage no longer provides any clear indicator of transport.

The argument for horse milk at Botai may also need to be revisited. With new methods, ancient proteins directly linked to milk products can be identified in ceramic residues or dental plaque from human teeth, a far more direct line of inquiry than isotope values. So far, efforts to identify these proteins from Botai ceramics and dental calculus have failed to yield evidence of any dairy consumption, much less dairy from horse milk.[60] In fact, a recent study of milk proteins from the teeth of early herders in nearby areas of western Eurasia suggests that people consumed only sheep milk at this time.[61] In light of this new information, it now seems exceedingly unlikely that Botai people consumed horse milk. Instead, the isotope patterns found on Botai ceramic potsherds are likely to have an alternative cause that may be revealed with better comparative reference samples.

In many ways, a hunting-only model for Botai provides a far more elegant explanation for the available archaeological evidence. The apparently sedentary life of people at Botai, who built a village of pit houses and appear to have lacked any other domestic animals or animal transport, is also very difficult to reconcile with the regular, long-ranging movements necessary to avoid overgrazing and disease among specialized domestic horse herders. Context is everything; components of the Botai record such as post-hole arrangements or decayed vegetative matter once thought to be dung layers or corrals can now be seen as innocuous remnants of hunter-gatherer structures.

Just as many Paleolithic horse-hunting spots were chosen for their location near a water source/crossing or their position on a migration corridor, Botai is located near a riverbank in an area that could have been a suitable mass harvesting locality in the summer months. Age and sex profiles showing a roughly equal proportion of healthy adult male and female horses between three and eight years, artwork made using horse bones, and even an active industry of tool making using horse bones all link Botai comfortably with Paleolithic hunting traditions. Like Deriyevka before it, Botai now seems most parsimoniously understood as a site where people conducted sophisticated and specialized mass harvesting of wild horses. The Botai hunters continued the final chapter of a Paleolithic legacy that once stretched across Eurasia and the Americas.

SUMMARY

In discussing this large and complex body of evidence for late Holocene horse use in Eurasia, we see that identifying early horse domestication is particularly difficult; after all, domestication itself is a continuum of different types of relationships between humans and horses. Some candidates for early intensive domestication relationships, such as Deriyevka and Botai, have shown promise, but advances in archaeological science have burst their bubble as evidence of raising or riding horses before the 2nd millennium BCE. How, then, can we explain the first emergence of horse domestication?

BEAT TWO

THE CART

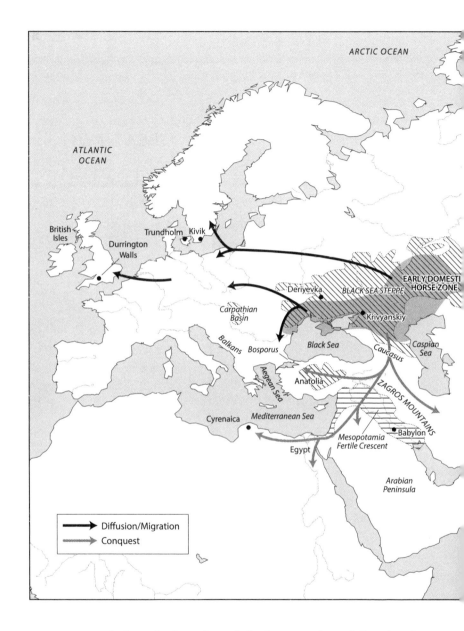

MAP 2. The domestication and spread of early transport horses in Eurasia and North Africa. Map by Bill Nelson.

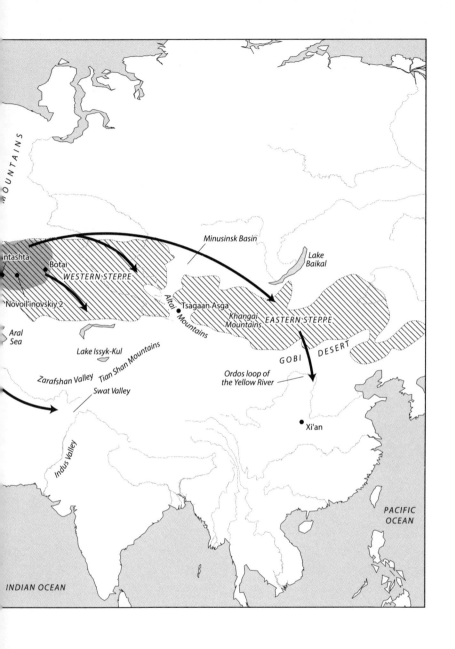

4 WHEELS

Around the end of the 4th millennium BCE, a major shakeup was taking place in the relationship between humans and large animals in western Eurasia. With agricultural communities crystallizing into urban centers and prime farmland at a premium, ancient farmers developed technological innovations that made life and agricultural labor more efficient. One of the most important of such innovations was wheeled animal traction, which began in western Asia or eastern Europe with the biggest and strongest of the early domesticates, cattle (*Bos taurus*).

The archaeological record is still not fully resolved as to precisely when or where the very first cattle transport began. However, pathological changes to the lower limb bones of archaeological cattle from the Balkans, in southeastern Europe, suggest that cattle might have been used for plow and traction work by as early as ca. 6100–4500 BCE.[1] The first type of animal-powered vehicle used to pull humans was probably a simple sledge, derived from earlier agricultural plows.[2] The augmentation of sledges with a rolling wheel (whether inspired from the pottery wheel or independently invented) turned them into all-terrain vehicles, and they became a big hit. By the 3rd and 4th millennia BCE, wagons could be found in cultures across the Fertile Crescent and eastern Europe.[3]

Wheeled carts, drawn by a pair of domestic cattle, could be used to haul large loads of people or goods efficiently over long distances. Wheels proved especially useful in the mountains and the steppe, where they opened a new realm of possibilities for mobility and migration. Cattle remains from sites of the Kura-Araxes culture in the southern Caucasus Mountains dating to the early 3rd millennium BCE show skeletal pathologies of the hip and lower limbs suggestive of their use in transport for plows or cart pulling.[4] Wheeled wagons spread across the Caucasus Mountains, where they first appear among the rich kurgan burials of the Maikop culture.[5] The presence of wheeled vehicles had an important impact on Maikop subsistence; while they made use of both agricultural grains and wild resources, the Maikop invested less in their residential architecture and raised few pigs, which are often associated with a sedentary lifestyle.[6] Concurrently, wagons or wagon parts also appeared in other areas across Eastern Europe and the Black Sea steppe, including those of the Yamnaya culture.

These early cattle carts were relatively simple in construction and used a control system specifically designed for cattle anatomy. Maikop kurgans contain unique copper objects, which, as discussed in chapter 3, were once thought to be an early kind of bridle component used for horse riding.[7] However, recent archaeological investigation has revealed instances of these artifacts in situ, showing them to be nose rings for cattle under harness to which control reins would have been tied(see plate 5).[8] The wagons themselves drawn by these early transport cattle were heavy four-wheeled carts, employing solid wheels cut from wooden planks.[9] A heavy yoke or crosspiece transferred the weight of the cart and its contents from the central bar to the strong neck and forelimbs of the cattle.

Cattle transport expanded the choices available to herders at the margins of colder and drier steppe environments, where grass was plentiful but the margins for crop agriculture were thin. A herder on wheels could draw heavy loads over long distances and pack up camp or relocate easily. The early wagoneers of northern Eurasia sometimes moved tremendous distances. With the power of the wheel, Yamnaya people or their close genetic relatives spread northwest from the Black Sea steppe and into central Europe. Their descendants became the Corded Ware culture and disseminated Indo-European ancestry and languages across a huge portion of the continent.[10] Yamnaya migrants also followed the grassy belt of the steppes eastward as far as south-central Siberia, where they became known as Afanasievo culture and raised domestic live-stock identified in the Russian Altai by around 3300–2900 BCE.[11] In fact, our own collaborative work with a team of archaeologists and geneticists now shows that this freewheeling migration reached all the way into the Mongolian steppe. Using the full recon-structed genome from two ancient Mongolian burials at the site of Shatar Chuluu in the Khangai Mountains of central Mongolia, we found that two ancient peoples dating to around 3000 BCE were of Yamnaya-like ancestry.[12]

In a matter of centuries, then, these herders traveled a distance of more than four thousand kilometers from the Black Sea steppe, leapfrogging across rich river valleys and the forest-steppe margins over Inner Asia into the heart of the Mongolian plateau. These early herders buried what may be deconstructed wagons in their tombs, along with remains of cattle, sheep and/or goat, and domes-tic dogs.[13] Tiny milk proteins recovered from inside dental plaque of the herders from Shatar Chuluu show us that Mongolian Afanasievo used dairy products as well, from sheep and perhaps from cattle.[14]

Over the decades, many scholars have sought to link these extraordinary Yamnaya migrations to the first domestication of the horse. Most recently, researchers noted that Yamnaya burials show skeletal problems in the lower back and pelvis/hip region, similar to those found in later populations of mounted horseback riders, leading to speculation that the Yamnaya were engaged in mounted riding, too, as early as around 3000 BCE or before.[15] But if this is true, where are the horses?

Quite tellingly, no securely dated or identified horse bones, horse milk proteins, or any other evidence for horse domestication can yet be connected to Afanasievo culture in Mongolia or neighboring regions. Early Yamnaya burials contain no artifacts linked with horse equipment, no iconographic indicators of riding, and no horse skeletons with pathologies linked with horse transport (in fact, almost no horse skeletons whatsoever). Almost certainly, the skeletal pathologies found in humans from early Yamnaya contexts were caused by long hours spent driving the cattle carts that are found in burials and iconography linked with this period, which would also place the driver in a seated position for long bumpy rides that could cause the same sort of pathology to the lower back and limbs. The archaeological record shows clearly that wheeled cattle transport, not horses, stimulated the transcontinental herder movements of the 4th millennium BCE.

Although the first wheeled cattle transportation gave cultures like the Maikop and Yamnaya the power to move, these lumbering carts were slow and clunky, limiting the distances that could be traveled over short time frames, like when moving herds between

FIGURE 5. Afanasievo herders set up camp at Shatar Chuluu, in central Mongolia, along with their early domestic animals. Drawing by Barbara Morrison.

pastures. In the steppes and deserts of Inner Asia, much of the rain that falls comes only seasonally, meaning that pastures can quickly be overgrazed by herders sticking to one spot. When paired with frigid winters, these conditions make sedentary herding or ranching nearly impossible in many areas.

Perhaps because of these limitations, early steppe pastoralists kept largely to the wetter northern zones and well-watered margins of the forest-steppe, where higher levels of rainfall could sustain herds, supplementing their diet with many other strategies (including hunting, fishing, gathering, and farming). Mongolia is one of the world's driest grasslands, receiving nearly all its annual moisture during a short rainy burst in mid-summer. However, Mongolia's high mountain zones (the Altai in the west and the

Khangai in central Mongolia) draw a slightly higher and more stable level of rainfall. This mountain rainfall is supplemented by perennial alpine ice accumulations that melt in the summer, providing freshwater and productive summer mountain pastures that support stable pastoral activities year-round.[16] These mountain zones served as a safe and habitable oasis for Afanasievo herders, and archaeological evidence for this culture is found only along the mountain margins of the Altai and Khangai.[17] However, without horses, neither the Afanasievo nor the subsequent herding cultures that followed them seem to have made inroads beyond these mountain footholds during the 3rd or early 2nd millennia BCE.

THE FIRST EQUID TRANSPORT

As cattle carts unlocked more of the steppes for semi-mobile herders in Inner Asia, the first domestic equid was being recruited half a world away for use in transportation. This trailblazer was not the horse but the unassuming donkey (*Equus asinus*). After domestic plants and animals spread into the fertile Nile River corridor during the Neolithic transition, this area experienced important changes to society, economy, and culture. Large urban centers and complex polities emerged across the upper and lower stretches of the Nile. And as in other early agriculture societies, there was a lot of work to do.

A demand for agricultural labor in early dynastic Egypt may have been the motivating force that brought donkeys into the human sphere. Donkeys appear to have been first domesticated from a wild progenitor in northeast Africa sometime prior to 3000 BCE, although recent genomic estimates may suggest an even earlier date, closer to 5000 BCE.[18] The ancestor of these earliest donkeys was most likely the now extinct Nubian wild ass.[19] At the site of Abydos, in Egypt, donkey burials dating to around 3000 BCE show vertebral deformation, indicating that they were used for transport, either for carrying heavy loads or perhaps mounted riding, a form of transport also shown in some early Egyptian iconography.[20] Despite their use for meat in some contemporary populations, there is so far little evidence to date that these first domesticated donkeys were raised for meat, milk, or other products. Instead, they seem to have been specifically domesticated as a transport animal.

DONKEY TRANSPORT IN THE NEAR EAST

After their cameo in early Egypt, donkeys disseminated outward to east Africa, Asia, and Europe. Stable isotope values in animal bones and teeth help archaeologists trace ancient processes of movement, trade, and exchange. Such analysis of donkey remains from the early Near East suggests that donkeys were exchanged into areas of the Levant and the eastern Mediterranean during the early 3rd millennium BCE.[21] Along the way, donkey transport collided with the early tradition of cattle transport and wheeled vehicles.

Donkey transit offered many advantages in the warm deserts of western Asia. Asses were better adapted to work in high temperatures, needing less water than cattle. Although not as strong as cattle individually, donkeys were lighter, sleeker, and faster, and when hitched to a cart in teams, they could reach higher speeds than their bovine brethren. In Asia, donkeys surged in popularity and became a crucial aspect of the social and economic infrastructure of the Near East. Texts and images from this region during the

3rd millennium BCE chronicle donkey use in agricultural activities such as plowing, threshing, and packing, as well as in commerce and trade.[22]

EQUID TRANSPORT, ROUND 2

The synthesis of Asian wheeled vehicle technology with African donkeys awakened experimentation with other large mammals. In addition to the donkey, several other wild equid species were native to the mountains and deserts of western Asia. These included caballine wild horses, hemiones such as the European wild ass (*Equus hydruntinus*), and several subspecies of the Asian wild ass (*Equus hemionus*), including the Syrian subspecies (*E. hemionus hemippus*) and the Iranian subspecies (*E. hemionus onager*). These graceful creatures probably first evolved in the deserts of western Asia before their relatives colonized areas of Tibet and the grasslands of Mongolia during the middle Pleistocene.[23]

As a native taxon, the wild ass was particularly dressed for success in the Fertile Crescent. Although donkey-like in appearance, hemiones are sleeker and faster than their African cousins, and DNA analysis suggests that they are more closely related to horses than to donkeys or zebras.[24] Evolutionary proximity is paralleled in the hemione's astonishing speed, which is even greater than that of the donkey and nearly equal to the horse.

The textual record from ancient Mesopotamia suggests that following the introduction of the donkey, hemiones were also temporarily brought into the domestic sphere. Historical texts make reference to many different varieties of domestic equids in old Sumerian, an early language of the civilizations of southern Mesopotamia.[25] One animal identified in these texts is ANŠE.

LIBIR, which appears to have been crossbred with other equids to produce a hybrid known as ANŠE.BAR.AN.[26] Based on available inscriptions, it seems that ANŠE.LIBIR were at times used for agricultural activities and to pull wheeled vehicles; hybrids were favored for pulling wheeled vehicles and even riding as early as the Early Dynastic IIIb period (the second half of the 3rd millennium BCE).[27]

Near Eastern iconographic depictions from the 3rd millennium BCE provide visual snapshots of early equid transport in action. Perhaps the most famous such example is the ornamented box dating to the 26th–25th centuries BCE, commonly known as the Standard of Ur. On the side of the box, a colorful mosaic scene shows soldiers going to war in four-wheeled carts, each drawn by a four-animal team of short-haired equids. In the lower panels, the carts brutally trample their unlucky enemy under rumbling hooves, making clear that their purpose was for war.

But what kind of animal was ANŠE.LIBIR? Basing their inferences on assumptions about the Yamnaya and the archaeological record for horse domestication, many scholars have presumed that this equid was an early domestic horse. If so, ANŠE.BAR.AN would be the animal we know today as a mule or a hinny, the hybridized offspring of a domestic horse and donkey. However, Sumerian texts have different terms entirely for the horse and the mule, and neither word appears in textual records until *after* the start of the 2nd millennium BCE.[28] A more recent comprehensive summary by Recht identifies ANŠE.LIBIR as a domestic donkey.[29]

There is a strong likelihood that most if not all equids depicted in transport from early Near Eastern iconography represent donkey, hemione, or hybrids and not domestic horses as is commonly assumed.[30] The Asian wild ass is smaller than the horse and has some qualitative differences in its tail and mane with domestic horses, but it also has short, horse-like ears, which muddy the waters for distinguishing the two.[31] Scholars who attempt to discriminate between equids in artistic imagery usually end up relying, therefore, on small variations in the depiction of tails and mane, such as the inferred presence of absence of a tufted tail (common in asses/hemiones) and forelocks (mane hair extending to the forehead on horses). However, the level of abstraction used in most 3rd-millennium BCE iconography prevents much meaningful or reliable inference on subtle details like this. In fact, the most commonly cited example of an early horse rider (a baked-clay plaque dated to the early 2nd millennium BCE from southern Iraq) was initially identified as a large wolfhound![32] Among images shown in high anatomical detail, only a small handful of objects in the region seem to show caballine horses, and even these typically seem to depict the wild *E. przewalskii*.[33]

For the most part, the greater the anatomical detail, the greater the clarity that a donkey or wild ass, rather than a horse, is depicted in images of ancient Mesopotamian transport equids. Most Near Eastern equid transport images from the 3rd millennium show sleek, hairless animals with an erect mane, often with a clearly tufted tail. These characteristics occur naturally in the hemione and donkey but not the horse. In many cases, transport equids are also decorated with stripes along the animal's back.[34] This unique color pattern is associated with ancestral donkeys.[35]

The most relevant information in ancient equid imagery may actually be the humans shown riding them. Both donkeys and hemiones lack a high withers, or shoulder area, distinguishing them from their caballine cousins. This anatomical peculiarity means that human riders sitting astride must place their weight far to the back of the animal, where it can be more comfortably distributed over the rear limbs, a posture commonly referred to as the "donkey seat" by archaeologists and historians.[36] For those looking for evidence of horse riding, the perplexing ubiquity of the donkey seat in ancient Near Eastern imagery is sometimes chalked up to poor familiarity with horse anatomy. However, a far simpler explanation is that these earliest riders were all riding astride donkeys, hemiones, or their hybrids.

THE ARCHAEOZOOLOGICAL RECORD

The use of donkeys and hemiones, and not horses, for transport in the ancient Near East is also borne out by the archaeozoological record. Parsing equid taxa based on bones alone is a logistical challenge, which has historically forced scholars to compare subtle differences in the size and shape of teeth or particular skeletal elements against small modern reference groups. Nonetheless, even using this error-prone approach, precious few horses have ever been identified at archaeological sites in most areas of the Near East prior to 2000 BCE, except in the mountain zones of Turkey and Iran where wild taxa also lived.[37] Among those caballine horses that have been identified predating the 2nd millennium BCE, none have yielded osteological evidence of use in transport. The oldest archaeozoological evidence for horse transport south of the steppe comes from the early 2nd millennium BCE site of Tal-e Malyan in modern day Iraq, where a caballine horse jaw exhibited bit wear showing that it was bridled and used in transport.[38]

The question of how equids were used in the ancient Near East may finally be resolved by archaeozoology, paired with new genomic techniques. Burials of small, ass-like equids are well known from 3rd-millennium BCE sites across the Levant and Mesopotamia.[39] At one site in western Iran, known as Godin Tepe, comparison of the internal structure of leg bones morphologically identified as hemiones (*E. hemionus*) recovered from site components dated to ca. 2000 BCE suggested that these animals were used for work.[40] At the Syrian site of Umm el-Marra, dated to the middle of the 3rd millennium BCE, archaeologists recovered twenty-six equid skeletons from both middens and burials (see plate 6). Careful morphological comparison of these mostly intact skeletons identified them all as donkeys, hemiones, or hybrids.[41] In 2021, researchers conducted genomic analysis of Umm el-Marra equids, revealing that they were the offspring of a male Syrian wild ass and a female domestic donkey.[42] This discovery demonstrates without a doubt that a hemione—specifically, the smaller Syrian subspecies—was bred with donkeys to produce ANŠE.BAR.AN. These hybrids would likely have been stronger and faster than donkeys and more docile than wild hemiones, making them more tractable in transport. The likely sterility of the hybrids, however, would have meant that regular crossbreeding was necessary.[43]

EARLY HEMIONE AND DONKEY TRANSPORT TECHNOLOGY

As donkeys and hemiones rose in cultural significance, transport technology had to be adapted to the challenges of equid transit. The first donkey-drawn vehicles, like those on the Standard of Ur, were lumbering affairs—heavy vehicles with fixed axles and solid wooden disk wheels formed from composite planks. The thick wheels of these early carts were sometimes studded with metal or

FIGURE 6. The heavy battle cars of the mid-1st millennium BCE and the lighter, silly "straddle car" (*both top*) show different stages in the adaptation of heavy wheeled vehicles to equid speed in the ancient Near East. An early rider astride a hemione (*bottom*) is shown using a whip, girth strap, and rein-and-ring control and seated in the "donkey seat" to the rear of the animal. Drawing by Barbara Morrison.

bracketed with metal tires.[44] For a time, the vehicles drawn by both cattle and equids seemed to be of nearly identical design.[45]

Before long, however, these plodding, heavy-duty transports were adapted for the anatomy and speed of asses and hemiones. Cattle yokes were adapted to the equine neck via a yoke saddle, a V-shaped tree that distributed the weight of the yoke and kept it situated forward of the withers/shoulders.[46] Four-wheeled tanks gave way to two-wheeled vehicles, like the enigmatic straddle car, that carried riders in new ways.[47] This light conveyance retained the basic technological components of the wagon (a central pole, heavy wheels, and axle), but did away with two wheels and the heavy platform, adding handlebars and a sort of bicycle seat to host a single rider. The resulting contraption looked something like an early Mesopotamian Segway. The straddle car was eventually replaced by the platform car, a sort of proto-chariot employing a screened platform but still using solid-disk wheels.[48]

Despite some indirect suggestion of an earlier date for this technology, the oldest bits recovered from a dated archaeological context postdate the start of the 2nd millennium BCE.[49] With no bridles or bits yet invented, early equids were apparently controlled with simple nose and lip rings, in the same fashion used in cattle carts. This ring provided the driver with very little directional control over each individual animal and probably functioned primarily as a braking device.[50] However, with a team of two or four animals under rein, more sophisticated navigational maneuvers and turns would probably still have been feasible by adjusting tension to individual animals.[51] Often, the reins were kept organized through the use of separators, or rein rings, attached to the pole, which helped prevent the lines from crossing or tangling.[52]

This system of control left a clear and identifiable impact on the skeleton of early transport equids. At Umm el-Marra, a hard material appears to have worn away the thin alveolar bone and hard protective enamel coating above the gumline for front upper incisors of the hybrid teams, where the tooth surface would normally not be exposed to damage (see plate 6). The only viable explanation for this pattern is the prolonged use of a metal lip ring like those seen on early iconography. The asymmetric patterning of the wear probably reflects the animal's role on the left side of a two- or four-equid draft team. This find helps us understand the ring-rein system in better detail; the anatomical location of the el-Marra wear implies that the ring was likely passed through the flesh of the upper lip rather than the nose.

When ridden, both donkeys and wild asses were apparently also controlled with a rein-and-ring setup. This form of early equid riding was a bit risky. In addition to the clunky nature of the lip-ring system, riders had no saddle and appear to have been secured only with a girth strap looped under the animal's chest. At the time, riders probably mounted their animals only for less dangerous situations, or for specific tasks that prioritized speed, such as serving as messengers or scouts, goading their animals forward with the use of a whip.[53] In most cases during the 3rd millennium BCE, wheeled vehicles remained the most stable, useful, and reliable form of transit.

SUMMARY

A wide array of linguistic, iconographic, and archaeozoological/ biomolecular data thus converge to demonstrate that donkeys and hemiones, not horses, were used for wheeled transport and mounted riding during the 3rd millennium BCE. Beginning first with domestic donkeys that likely arrived in the Near East from northeast Africa, residents of this region made progressive innovations in wagon and harness technology designed for cattle traction, adapting them to a faster, lighter, chariot-like vehicle that was still drawn by a simple nose-ring and heavier disk wheels. The speed of E. hemionus encouraged experimentation with domestication and crossbreeding with donkeys, producing a multispecies array of transport equids that were sometimes also used for mounted riding.

With faster and more effective vehicles, Near Eastern societies could plow the fields, thresh and transport grain, and trade with greater efficiency than ever before. Equid messengers and long-distance communication would have helped to forge new connections with nearby regions like Central Asia, northern Africa, and the margins of Europe. Donkey and hemione carts proved useful in battle and helped facilitate the rise of early city-states, which could breed the desert-adapted animals effectively and easily. But in the cold grasslands to the north, a relationship between people and horses was emerging that would soon have implications that reached across the ancient world.

5 CHARIOTS

THE FIRST HORSEMANSHIP

With improved scientific data eroding earlier candidates like Deriyevka and Botai and no reliable evidence for horse use among the early Yamnaya or in the ancient Near East, where do we turn for insights into the earliest domestication of the horse?

Over the last several years, an unprecedented research project analyzed and dated hundreds of ancient horse bones from archaeological sites across Eurasia, seeking genomic clues to the origin of the first domestic horses. This work now shows beyond a doubt that the earliest ancestors of the domestic horse were a wild caballine taxon occupying the Black Sea region and nearby areas of eastern Europe and Anatolia.[1] These new DNA data tell us that sometime before the middle of the 3rd millennium BCE (ca. 2900–2600 BCE), the first direct ancestors of the domestic horse, known as the DOM2 lineage, appeared in archaeological sites of the very late Yamnaya culture. These early proto-horses initially occupied only a small area of the Pontic-Caspian steppes between the Dnieper and Don Rivers in modern-day Russia and Ukraine.[2] Although for many years, scholars had hypothesized that the ancestor of the domestic horse was a wild horse called a tarpan (also called *E. ferus ferus*) and known from historical observations in southwestern Russia, sequencing of tarpan specimens showed them to be a later hybrid between indigenous European wild horse lineages and the DOM2 domestic horse.[3]

Answers as to what brought horses under direct human control in the Black Sea region during the late 3rd millennium BCE are still scarce, but new scientific analyses from human burials dating to this time provide some clues. For many thousands of years prior, hunters in the Black Sea region already had detailed knowledge of *Equus* ecology and behavior, hunting them as an important part of their diet. In an analysis of sixteen early Yamnaya and Pontic-Caspian burials, a team of researchers led by scientist Shevan Wilkin (and including myself) used ancient milk proteins preserved in the teeth of ancient herders to show that the 4th millennium BCE saw the adoption of other domestic animals, including cattle, sheep, and goats, for dairy products.[4] When domestic livestock and animal transport technology entered this region, the Yamnaya were perfectly positioned to try something new with the wild equids of the Black Sea steppe.

Among the early Yamnaya samples we studied for ancient dairy products, we also identified two teeth with dietary proteins unique to horses, most likely from *E. caballus*. Hailing from the site of Krivyanskiy on the banks of Russia's Don River, these teeth now form what could be the oldest direct evidence of humans managing horses as a domestic animal.[5] However, until more corroborating evidence in found, these data may not be totally unimpeachable on their own. Radiocarbon dates from human remains can sometimes be biased. In particular, organisms eating a diet rich in fish or other marine resources can cause the incorporation of carbon with different isotope values, causing a date from the organism to appear older. After considering the possibility of dietary effects, we can be less confident that these first horse milk proteins date to the early 3rd millennium BCE or before.

If Yamnaya herders did experiment with horse milk, however, the overall importance of this contribution to their diet was probably negligible or nonexistent. Beyond the isolated evidence for horse milk found at Krivyanskiy, recent large-scale studies demonstrate that Yamnaya and other early pastoralists relied almost exclusively on milk from sheep until around 2800 BCE and thereafter incorporated primarily goat and cattle milk, but not horse milk, until the 1st millennium BCE.[6] So far, very little other archaeozoological data are available to understand the role horses might have played in terms of reproduction, meat production, transport, or any other activities associated with domestic horses. In fact, in most cases, it is unclear whether horse remains from 3rd-millennium BCE Black Sea sites originated from wild or domestic animals. Until further research can help sort out these important questions, it is difficult to quantify just how important the first horses were in the Pontic-Caspian economy.

THE ORIGIN OF HORSE TRANSPORT

Within a few centuries of their first appearance in the archaeological record, domestic horses became a transformative force that electrified the ancient world, from Africa to East Asia. At the core of this dramatic change was a single innovation: the horse-drawn chariot.

As donkey and hemione transport expanded across western Asia during the 3rd millennium BCE, the continent became increasingly connected through cultural exchange, conflict, and trade. After cattle carts were adapted to the donkey in Mesopotamia, other wild equids like the hemione were an obvious temptation for experimentation. But while donkeys and hemiones thrived in the warm desert zones, the colder climates of the steppe were relatively unsuitable for these faster mammals. As a result, donkey and hemione carts failed to spread significantly northward into the Black Sea region.

Horses were stronger and faster than any existing transport animal, and unlike the donkey and the hemione, horses were already adapted to the northern steppe, making them an especially tempting target for steppe herders already familiar with cattle and equid transport. However, at least two major barriers stood in the way of using horses for this purpose: their behavior and their anatomy. Horses are prey animals, with anatomical, social, and cognitive systems designed to avoid dangerous predators—like us. For the last half-million years, the most significant predator of the horse has been *Homo sapiens*, and close physical proximity sets off a fight-or-flight response from a wild animal that can be exceedingly dangerous (as anyone who has attended a rodeo can tell you). Because of the horse's great strength and speed, navigating these behavioral challenges was probably one of the foremost barriers to early horse transport. Recent genomic research corroborates this idea, showing that the strongest selective pressures exerted on early *E. caballus* by humans focused on reducing their aggression.[7]

OVERCOMING AGGRESSION

Using a cart helped overcome some of the behavioral challenges posed by early horses. For a wild animal, few activities induce more panic than being directly mounted by a rider, where the human lies outside its field of vision.[8] This panic is mitigated to some degree by a cart arrangement, where the driver is situated farther to the rear and not in direct contact with the animal. Perhaps this flight reflex was even useful in a cart, serving as motivation for the horse to pull forward. Because horses are herd animals, the presence of additional animals in a driving team also has a soothing effect.[9] A cart also minimizes physical risk to the human; a fall from a vehicle platform is less dire than being thrown from an animal's back.

For wild equids with no behavioral adaptations to humans, these factors and others conspire to make carts inherently easier and safer than mounted riding. This assertion may seem counterintuitive to those familiar with modern domestic animals, which can often be ridden bareback or with a minimum of equipment. Why bother with such extensive equipment when one can simply hop on and ride astride?

THE CART BEFORE THE HORSE

The advantages of carts with aggressive or unpredictable animals are nicely illustrated by a modern story about the zebra. From the 17th through the 20th century, the United States and many Western powers were busily engaged in colonial exploitation of Africa. At this time, horses were the backbone of Western transport, industry, and military power. This was a problem in tropical regions of Africa, where imported horses experienced extremely high mortality, quickly perishing from insect-borne diseases such as trypanosomiasis and African horse sickness.[10]

But wild zebras, as equids indigenous to the African continent, fared much better against many of these disease barriers. While imported horses died in droves, some ambitious Western powers sought to domesticate the zebra for transport, beginning as early as the mid-17th century with Dutch colonial authorities in South Africa.[11] In the United States, similar efforts began in the early 20th century. One ornery zebra named Dan, given to President Theodore Roosevelt by the king of Abyssinia, kickstarted an ill-fated government breeding program seeking to produce a zebra with a manageable temperament.[12] Such programs had little success. As one South African newspaper sadly proclaimed, the zebra's temperament rendered them "wholly beyond the government of man."[13]

While they never found regular use for riding, some zebras *did* experience a brief moment of popularity in pulling carts and carriages. In a carriage harness, even these flighty, highly aggressive animals could be used somewhat safely. For a brief while, novelty zebra carriages sprang up from South Africa to New Zealand and even New York.

Other historical examples abound of wild, aggressive equids being effectively harnessed into cart transport. Based on the account of a former carriage driver on the Overland Stage linking Kansas to Colorado during the 1860s, stage drivers in the 19th-century American West used to regularly hitch up wild, unbroken mustangs, either out of boredom or because they were faster and more spirited than other horses when traveling on a tight schedule over level terrain.[14] On the overland wagon lines, sometimes even wild bison were hitched to ox carts.[15] These historical examples illustrate the very real advantages of wheeled vehicles in adapting undomesticated animals to human transport needs during the earliest stages of horse domestication.

ADAPTING TO ANATOMY

A second obstacle to horse transport was posed by the unique physical traits of caballine horses. Unlike donkeys and asses, which are ridden with the "donkey seat" that places the rider's weight above the robust pelvic girdle, sitting astride a horse puts tremendous weight on an unsupported stretch of the animal's spinal column between the shoulders and the rear legs. Therefore, developing a reliable form of horse transport also required developing a system that was anatomically stable for the horse's vulnerable mid-spine area. Genetic analyses suggest that the health of this vulnerable spinal column may have been a particularly significant focus for the earliest horse breeders.[16]

Although cart transport was one possible solution to this issue, the solid wooden wheels used in donkey/hemione carts were also heavy and poorly adapted for high speeds. Moreover, the simple rein-and-ring system used for cattle, donkeys, and hemiones provided a very simple level of control that might have made even horse carts too dangerous for regular use. To harness horsepower for the first time, technological innovation was needed.

THE CHARIOT

Building on earlier technology, the herders of western Asia worked out a brilliant set of tools for harnessing the first horses. Sometime after around 2400 BCE, a late outgrowth of the Yamnaya culture known as the Catacomb culture began experimenting with their own version of the two-wheeled vehicle tradition on the northern edge of the Caucasus and the Black Sea steppe,

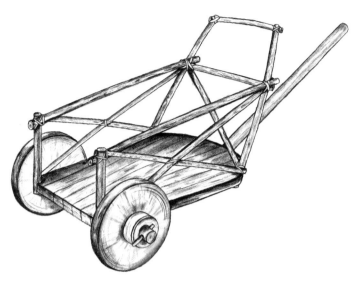

FIGURE 7. Artistic representation of a two-wheeled cart from the Tyagunova Mogila Cemetery (Ukraine). The image is based on a 3D CAD model reconstructed after a published scaled filed drawing from S. Pustovalov's "The 'Tjagunova Mogila' Burial Mound and the Problem of Wheeled Transport." Drawing by D. Chechushkova.

producing lighter carts with small, single-piece disk wheels.[17] While Catacomb sites do not have any unambiguous evidence for horse transport, archaeozoological data show that this culture also engaged in consumption and ritual burial of horse remains at several sites.[18] From light, two-wheeled vehicles to animal pastoralism and familiarity with wild horses, all the key ingredients for early horse transport were present in the Black Sea steppe at the end of the 3rd millennium BCE.

The first clear evidence for the use of domestic horses in transport comes from sites of what is known as the Sintashta culture, located near the northwest border between Kazakhstan and

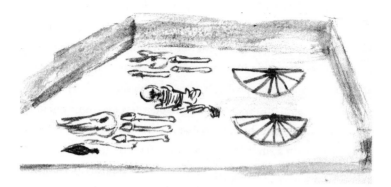

FIGURE 8. Artist's sketch of the partial horse and chariot burials at Sintashta, in the Trans-Ural region of Russia. Drawing by Barbara Morrison.

Russia.[19] In the very earliest decades of the 2nd millennium, Sintashta burials begin to include evidence of horses, horse equipment, and chariots. In a given burial, remains of horse heads and lower limbs are typically found in pairs along with soil impressions that show they once pulled a light two-wheeled cart.[20]

Sintashta burials have produced the first archaeological evidence for two crucial technological innovations that helped adapt early wheeled vehicles to the behavioral and anatomical peculiarities of the horse. Replacing the earlier lip-ring system, Sintashta horsemen invented the very first bridle and bit. This critical device replaced the rudimentary lip ring of earlier donkey/hemione systems with a mouthpiece, secured to the animal's head on either side by two lateral cheekpieces affixed to a dual pair of reins. These rigid, often spiked plates aligned on either side of the mouth. When the reins were pulled on one side, the cheekpiece on the opposite side of the mouth would be pressed into the tissues of the horse's face, compelling the animal mechanically to turn in the desired direction and providing a much greater level of lateral control. These first Sintashta mouthpieces were made of soft organic material, which could be broken or chewed through.[21] However, with the safety in numbers of a two-animal chariot team, such occasional equipment failure was probably not catastrophic.

The Sintashta bridle also incorporated a new and more effective system of braking, the bridle noseband. For donkeys and hemiones, a nose ring with a single rein had functioned for both braking and turning. In the new horse bridle, these two signals could now be parsed: a band of material stretching from the cheekpieces across the animal's nose would exert a downward force on the horse's head when both reins were pulled, forcing the animal to slow. Together, these two new features provided drivers with nuanced control needed to steer the ornery *E. caballus.*

The second key innovation found at Sintashta was the spoke. Although solid plank wheels were sufficient for traveling at lower speeds, the speed and strength of horses necessitated a lighter and more robust wheel design. The spoked wheel, emerging from long-established expertise in bentwood carpentry by forest-steppe cultures, provided improved stability at high speeds.[22] Controlled with this new bridle and bit system, these light and stable horse chariots were faster than any existing cart-based transit system yet known to human societies.

THE EMERGENCE OF HORSE PASTORALISM

The chariot revolutionized pastoral life in the grasslands of Inner Asia. Before the horse, most cultures occupying the margins of the steppe zone actually lived in a circumscribed area year-round, practicing some combination of farming, herding, and hunting.[23]

In areas that were not productive for farming and that were susceptible to overgrazing, freewheeling horse-borne herders could raise larger numbers of more mobile, arid-adapted livestock like sheep and goat and move them to pasture across wider distances. With increased efficiency in pastoral pursuits, these early charioteers became less dependent on wild resources. At Sintashta and sites of subsequent forest-steppe cultures of southern Russia, including the Srubnaya or Timber Grave culture (17th–14th centuries BCE), wild animals declined drastically in significance, constituting only a minor portion of dietary assemblages during the first half of the 2nd millennium BCE.[24]

Horses themselves became far more useful as livestock, too. While large herds of free-grazing horses are nigh impossible to control on foot, horse transport allowed them to be effectively raised en masse, and horses became an important source of dietary meat. Based on the bones found in archaeological sites, horses began to account for up to 15 percent or more of dietary assemblages for many early steppe horse herders across southern Russia and northern Kazakhstan.[25] Sites from this region also begin to show clear human control of horse reproduction, including a high frequency of young males and older animals.

Addition of more horses into the pastoral economy probably increased the overall resilience of multispecies herds during the tough steppe winters. With their large hooves, horses are able to dig through icy snow crusts that can form in winter, meaning that they are able to expose grasses for grazing in areas where smaller livestock might otherwise starve during an icy winter.[26]

Following the first horse transport, secondary products from horses, such as milk, also became an important part of steppe life. Like many populations around the world, many Asian populations are lactose intolerant as adults. Yet societies that kept these early horses found a solution—the fermentation of horse milk. Fermentation breaks down sugars and improves milk digestibility, thereby helping provide key nutrients that diversify the otherwise protein-heavy diet of a pastoralist. Toward the end of the 2nd millennium BCE, horse milk seems to have been a core component of diets from Mongolia to Ukraine.[27] Bolstered by horse herding and transport, pastoralists and horses spread further eastward into Siberia and southward into the open grasslands of Kazakhstan and Central Asia.

USE OF CHARIOTS IN STEPPE SOCIETIES

What did these first horse-drawn chariots actually look like? In classical history, the chariot is often conceptualized as an elite conveyance, used only for warfare, racing, or the whims of the social elite. To some scholars, a true chariot is even explicitly defined as only those two-wheeled carts with specific battle capabilities, such as a platform wide enough to support mobile archery.[28] However, such visions of the chariot are probably a poor match for the vehicles used in the steppe during the 2nd millennium BCE. Although no complete steppe chariots have been recovered, the few intact chariots known from 2nd-millennium BCE contexts, such as those recovered in Tutankhamun's tomb in ancient Egypt, show that chariots were made from raw materials that were locally abundant in the steppe, like tamarisk and rawhide.[29] In all likelihood, these vehicles were light, locally produced, utilitarian affairs designed for herding life in the grasslands.

Some more direct insights into the social role of the steppe chariot can be gleaned from Inner Asian rock art.[30] In the Altai

Mountains, which form the imposing boundary between Mongolia and the western steppe, researcher Esther Jacobson-Tepfer and her colleagues have documented an immense array of ancient petroglyphs, including many Bronze Age scenes depicting horses and chariots. In these Altai panels, chariots are rarely, if ever, represented as part of battle scenes.[31] Instead, chariots are used in hunting deer or big game or appear in large, peaceful scenes with an abundance of animals.[32] Some panels could even represent chariots being used to guide domestic animal herds. In all cases, Central Asian rock art shows these ancient vehicles to be small and simple in construction, with a central pole, platform, and wheels and sometimes a box with a driver and reins.[33] Despite their functional role in steppe life, these first chariots were obviously a source of major social prestige, figuring prominently in funerary offerings, and in some areas, such as Sintashta, a rise in weaponry found in burials and an increase in fortified settlements points to their use in combat.[34] But these speedy, light carts were working vehicles, more like the Ford F-150 than the Rolls-Royce of 2nd-millennium BCE steppe societies.

THE ARCHAEOZOOLOGY OF EARLY CHARIOTS

Our best direct archaeological evidence as to how these first transport horses were used comes, once again, from archaeozoology. The most carefully studied horse remains from early 2nd-millennium BCE contexts are those from the site of Novoil'inovskiy 2, located just east of Sintashta on the banks of the Tobol River in western Siberia along the Russian-Kazakh border.[35] At this site, a pair of horses belonging to the Petrovka culture (closely linked to Sintashta) was radiocarbon dated to ca. 1800 BCE. The site's horses were buried along with bridle cheekpieces and bronze weaponry.

While no chariot remains were recovered from the site, careful osteological analysis of these skeletons by researcher Igor Chechushkov indicates that the team was bridled and used for transport. Recognizable depressions on the bones of the nose point to the prolonged use of a bridle noseband, while the limited wear to the front margins of the cheekteeth row suggests that the horses were controlled with a soft organic bit, like those found at Sintashta and other early steppe horse burials. The horses at Novoil'inovskiy 2 also show grooves on the premaxilla that are caused by heavy exertion.[36] Although they could have been ridden as well, if the pair of horses in the burial were indeed a chariot team, the male-female pairing might suggest that the male was castrated; pairing an "uncut" stallion with a reproductive mare creates a behavioral disaster.

THE SPREAD OF THE CHARIOT ACROSS THE STEPPE

Spurred by the power of the chariot, these early domestic horses radiated rapidly outward from their homeland in the western steppes and into Europe, Asia, and even northern Africa. Beginning around 2000 BCE, chariot horses rapidly dispersed over a wide range of the Eurasian continent over a period of only two to three hundred years.[37] Chariot burials and paired horse burials appeared in the Alakul culture horizon of the Trans-Ural region and the Petrovka culture in northern Kazakhstan.[38] By the middle of the 2nd millennium BCE, they had spread east to the Russian Altai; horses emerged in Fedorovo culture burials in southern Siberia and eastern Kazakhstan by around 1600 BCE and in sites of the Karasuk culture in the Minusinsk Basin by around 1500 BCE.[39] Domestic horses moved southward into the great grasslands of Kazakhstan and Central Asia and by the first half of the 2nd millennium BCE were found deep in the heart of the Tian Shan at Aygarzhal 2 in central Kyrgyzstan (ca. 1900–1400 BCE).[40] Steppe-associated pottery styles and settlements of steppe-linked cultures spread as far south as the Zarafshan Valley of Tajikistan, where Sintashta-style cheekpieces have been recovered in archaeological contexts dated to the early 2nd millennium BCE.[41]

INTO EUROPE

As their utility became apparent in northern latitudes, horses and chariots proliferated outside of the steppes, too. To the west, domestic horses and chariots radiated into the grasslands and forests of central and western Europe. Beginning with the Abashevo culture, horses and chariots were transmitted into Romania and the Balkan peninsula, as evidenced by chariot harness equipment found in 2nd-millennium BCE burials.[42] By around 1950–1750 BCE, chariot burials reached western Russia and might have spread at an early date into northeastern Europe, if young caballine horses at sites of the Únětice culture in Poland and the Czech Republic indeed represent domestic specimens.[43] Within a few centuries, horses were taken across the Baltic Sea as far as southern Sweden, where chariots are depicted on wall panels in the royal tomb at Kivik around 1300 BCE.[44] By the 14th–12th centuries BCE, both horses and chariots had reached the British Isles, shown by directly radiocarbon-dated materials from the site of Durrington Walls.[45]

While horses moved across northern Eurasia through processes like migration, diffusion, and trade, the spread southward was decidedly bloodier. From their homeland in the Black Sea steppe, the first entry point of domestic horses into lower latitudes seems to have been the Caucasus Mountains, which separate Europe from Asia Minor by connecting the Black and Caspian Seas.

Linguistic memories of the first horses and chariots are encoded in the ancient languages of Mesopotamia. Beginning in the 20th century BCE, the earliest textual references to chariots or spoked wheels come from areas abutting the Caucasus region, including northern Mesopotamia, northwestern Syria, and eastern Anatolia.[46] The Sumerian word for domestic horse, ANŠE.KUR. RA, which appears for the first time in historical texts, roughly translates to "equid of the eastern mountains."[47] The word for chariot warriors in the Akkadian spoken in ancient north Mesopotamia is *maryannu*, apparently a fusion of an Indo-Iranian root word using a plural suffix from an Anatolian language, Hurrian.[48] Akkadian inscriptions from the 18th century BCE suggest that the word for "spoke" was a foreign loan word.[49] These inscriptions show that the horse and chariot were both new and foreign at the time and suggest that the Caucasus and Anatolia served as a staging ground for the southward spread of the domestic horse.

The idea that horses reached western Asia through the Caucasus is also supported by archaeological data, ranging from the technological to the biomolecular. Horse equipment first known in the steppes, such as bitted bridles, appear south of the Caucasus only during the 2nd millennium BCE. And while these first Near Eastern bits are sometimes cast in bronze, their design seems to have been ported over directly from an earlier organic steppe predecessor.[50] Recent genomic study of archaeological equid remains across Turkey show that only wild horse lineages unrelated to the early domestic (DOM2) lineage were present in the region before the 2nd millennium BCE.[51] These genomic data also support the idea that horses arrived into Anatolia via the Caucasus rather than the alternative paths, like the Bosporus Strait. Together, linguistic, archaeological, and genomic clues all seem to converge to indicate a southward incursion of horses into the Near East, out of the Caucasus, in the early decades of the 2nd millennium BCE.

As horses and chariot technology left the cool northern grasslands, those with horses and chariots found themselves with a tremendous technological and military advantage over their neighbors and quickly rose to the top of the geopolitical food chain. During the 17th century BCE, people of the Hittite Kingdom, spanning Anatolia and Syria, were among the earliest to use chariots in warfare in the Near East.[52] Horse chariots allowed the Hittites to effectively siege and block access to an enemy city, and were supremely useful as both a mobile platform for firing projectiles or for rapid transportation of troops.[53] The great kingdom of Old Babylon, which had dominated southern Mesopotamia for centuries, was defeated by the Hittites in the 16th century BCE, before being taken over by the Kassites, another group of horse-wielding charioteers from the Zagros Mountains. The Hurrians of eastern Anatolia established control over the northern Fertile Crescent and eastern Mediterranean in the kingdom of Mitanni.

While most of these early conquests were undertaken by Indo-European language speakers native to northern Mesopotamia and Anatolia, others were also able to leverage early access to horses and chariots to ascend to power. For example, the Hyksos people, thought be Levantine or Semitic in origin, invaded Egypt from the north in the 18th/17th century BCE, razing cities and temples.[54] In the process, the Hyksos probably introduced the domestic horse and chariot into northern Africa for the first time, as horse remains first appear in Egypt's archaeological record during this time.[55] After the 17th century BCE, horses became common in ancient North Africa, with the archaeozoological record showing clear evidence of their use in transport.[56] Horses then trickled across the margins of northern Arabia and Mediterranean Africa, reaching Cyrenaica in coastal Libya and probably much of the northern Sahara by the later parts of the 2nd millennium BCE.[57]

Outside of the cooler steppes where they could be bred in large numbers, horses and chariots were difficult to acquire and became a focal point of military power and social prestige (see plate 7). Although horses themselves were rarely included in funerary contexts, several mummies are known from Egyptian archaeological sites. One tomb dating to the mid-2nd millennium BCE contained a well-preserved mummy of an adult female horse belonging to the architect of the female pharaoh Hatshepsut, along with a cloth object that could be an early saddle pad or other unknown piece of horse tack.[58] The tomb of the young Egyptian pharaoh Tutankhamun, who died in the late 14th century BCE, included six dismantled and elaborately decorated chariots probably imported from Anatolia or Transcaucasia.[59] Tack and horse equipment interred with the king included special "blinkers" and tools to prevent the horses from becoming distracted during battle. Excavations at the palace of Rameses the Great revealed extensive royal stable areas with facilities for nearly five hundred animals, including special halls for grooming, quarters for horse grooms, and even urinals built into the floor of each stall to keep the stable dry.[60] In some areas, Rameses's stables were so well preserved that they retained the hoofprints of the horses on the ground surface.

Thus, even in areas where they were difficult to raise, careful maintenance of chariot forces became a key task for maintaining political authority in the Mediterranean during the late 2nd millennium BCE. Chariot design, too, became optimized for combat, shifting to a blueprint using a rear axle better suited for high speeds.[61] Increasing reliance on horses seems to have prompted a decline in the use of hemiones and donkeys, particularly in high-speed transport. Across the Near East, horse-donkey hybrids (what we now would call a mule) became increasingly popular for traction and packing, rendering the hemione-donkey BAR. AN hybrids obsolete.[62]

SOUTHEASTERN EUROPE

The horse and the chariot rippled across the Aegean to the islands and peninsula of the Peloponnese. During the late 15th century BCE, a period known as the Late Helladic III, horse remains first appear in Greek archaeological sites and burials. Sometimes, horses appear as isolated skulls or skeletons; other times, they appear alongside other domestic animals.[63] Images of chariots first appear on palace frescoes at Mycenae, which also began occupation during this period.[64] Based on horse equipment, the date of their first introduction could be slightly earlier; a recent excavation at the tidal islet of Mitrou, in central Greece, yielded a horse bridle component likely dated to the 16th century BCE, while across the Aegean, horse bridle cheekpieces are found in level VI at the citadel of Troy, dating to around 1750–1300 BCE.[65]

Did these first horses and chariots come from eastern Europe to the north or across the water from Anatolia to the east? Historian Robert Drews, in his *Coming of the Greeks*, hypothesizes that the rise of Mycenean Greece was directly prompted by the invasion of chariot-riding peoples from eastern Anatolia.[66] However, other archaeological data, such as bridle equipment, might instead link these early horse chariots to contact with steppe cultures to the north.[67] In a 2017 study, researchers studying ancient DNA preserved in human remains from ancient Greece identified a contribution of between 4 and 16 percent from a new population of Siberian/Eastern European origin, likely introduced via either the steppe or the Caucasus.[68] Whether linked to an incursion from the east or the north, horses and chariots probably played a meaningful role in the emergence of early Greek civilization.

CENTRAL AND SOUTH ASIA

As charioteers rewrote the political landscape of the Mediterranean, another conquest was underway far to the south, in present-day India and Pakistan. During the mid-3rd millennium BCE, the Indus Valley had thrived with the rise of the Harappan civilization but underwent upheaval during the middle of the 2nd millennium BCE. This same period also saw the first arrival of foreign languages and practices, including the Sanskrit language, a member of the Indo-European language family linked with northern latitudes. In the texts of the *Rig Veda*, an ancient collection of Sanskrit sacred hymns, textual references appear to describe the domestic horse and associated rituals, practices that mirror those known from 2nd-millennium BCE archaeological sites in northern Central Asia.[69] A consideration of these factors have led many scholars to attribute the host of social changes in South Asia during the mid-2nd millennium BCE to an incursion of northern charioteers.[70]

From an archaeological standpoint, this hypothesis has proven difficult to assess. Both horses and chariots are exceedingly rare in South Asian archaeological assemblages, and even when equid remains are recovered from archaeological sites, few have been reliably identified to taxon or directly dated. At the site of Sinauli, in northern India's Uttar Pradesh Province, archaeologists recently discovered two-wheeled carts made of wood and copper.[71] However, the wheels of these carts are not spoked, like early chariots, and are instead outfitted with solid disk wheels of the type popular in the pre-horse era.[72] Moreover, the site has yet to be scientifically dated, and no animal remains were found with the site, meaning that the animals used to pull this cart may not have been horses at all.

Horses are essentially absent from sites of the Harappan and Indus Valley culture in Afghanistan and western Pakistan until the 2nd millennium BCE, supporting the idea that these animals were absent until this period.[73] Equid bones have been recovered from archaeozoological assemblages from the Ghalegai II period, thought to date to around 1800 BCE, but confidence in the species designation, or precise calendrical age of these assemblages is low.[74] A faunal study of the Swat valley site of Kalako-ḍeray, dated to 16th century BCE, found no evidence of equid remains, although equid bones are found at sites thought to be contemporaneous in the valley.[75]

Seeking an archaeological signature for the charioteer incursion, some researchers have turned to a cultural horizon known as the Gandhara Grave complex. This archaeological grouping, found in the Swat River area of northern Pakistan, is defined largely on the basis of its megalithic tomb style, although similar megalithic grave traditions are also found elsewhere along the Himalayan front.[76] The Gandhara Grave complex was once thought to span a wide temporal range from the 3rd millennium BCE through the late 1st millennium BCE.[77] Gandhara cemeteries in the Swat valley have produced at least two horse burials, with pathological bone formation of the lower limbs demonstrating their use in transport.[78] However, these horses have still not been directly radiocarbon dated, and recent large-scale dating of other burials at Gandhara suggests that most actually postdate the 2nd millennium BCE, making them of little relevance to the hypothesis.[79] Therefore, on the basis of current evidence, it is difficult to say whether there is any direct association between the Gandhara Grave culture and any earlier chariot or horse conquests.

Other archaeological clues, however, give the charioteer incursion idea greater credence. Bridle artifacts from southern Turkmenistan suggest that horses and chariots were near the margins of South Asia by the early 2nd millennium BCE.[80] Most significantly, a recent large-scale human genomic study found evidence of increased gene flow between India, Iran, and the Central Asian/Kazakh steppe during the early 2nd millennium BCE.[81] This gene flow appears to have been primarily from male intruders, indicating that it was probably conquest, not wholesale migration, that introduced the new ancestry. In total, then, the archaeological data would seem to support the idea that in India and South Asia as well, horses enabled another transcontinental movement and conquest of southern latitudes by early charioteers during the 2nd millennium BCE.

SUMMARY

From England to Egypt and Siberia to South Asia, the initial domestication of horses was followed by a speedy dispersal across three continents. In the steppes, horses were deeply integrated into herding life. While early horse transport probably did impact the structure of steppe societies, increasing warfare and elevating successful families and individuals to a higher social status, the animal's most important impacts in the steppe were its influence on mobility and pastoral life.[82] In contrast, in other areas, including the Mediterranean and South Asia, chariot-borne warriors overturned systems of political control, replacing existing systems of authority (often based in religion) with new military power structures.[83] In these zones, chariots became powerful tools of the elite. Many, but not all, of these groups spoke Indo-European languages, and other groups, like the Semitic Hyksos, were also able to effectively translate their early access to horses and chariots into political and military might.

Demand for both horses and chariots increased on a continental scale, forging new links between the steppes and the agricultural civilizations of western Asia. The archaeological record shows that horses helped globalize the grasslands; even the first chariot burials in Sintashta include far-flung trade goods, such as lapis lazuli from the mountains of Afghanistan and bronze mirrors linked to Central Asia.[84] As horses became the key to power, formerly marginal grassland regions now held the keys to transport and military might. Steppe horse cultures expanded to the margins of the steppe and integrated with the societies at those margins. Metallurgical products also began to travel across vast distances across the continental interior, and domestic crops formerly constrained to one corner of the continent (such as broomcorn millet in East Asia and free-threshing wheat from the Iranian Plateau), spread to opposite ends of Eurasia.[85]

In this way, horses helped build the first inklings of a truly globalized world in the 2nd-millennium BCE, moving peoples, goods, ideas, languages, and organisms into areas that they had never been before. Charioteers toppled dynasties and built new ones, now laid on a foundation of horsemanship and access to horses. By the late 2nd millennium BCE, horses were ubiquitous across much of western Eurasia, stretching even to the margins of the great grasslands of Mongolia, where the human-horse partnership would change humanity forever.

BEAT THREE

THE RIDER

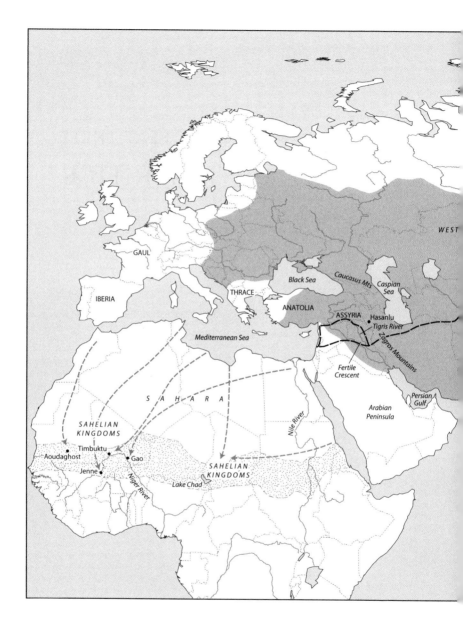

MAP 3. Horse-mediated linkages across the deserts and steppes of Eurasia and Africa. Map by Bill Nelson.

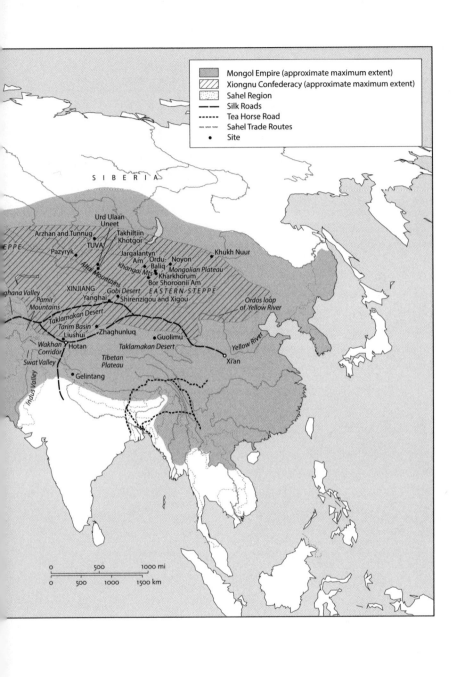

Legend:
- Mongol Empire (approximate maximum extent)
- Xiongnu Confederacy (approximate maximum extent)
- Sahel Region
- Silk Roads
- Tea Horse Road
- Sahel Trade Routes
- Site

SIBERIA

Urd Ulaan
Uneet
Arzhan and Tunnug · Takhiltiin
Khotgor
Pazyryk · TUVA
EPPE · Jargalantyn
Am
Khangai Mts · Ordu- Noyon · Khukh Nuur
Baliq
Kharkhorum · Mongolian Plateau
Bor Shoroonii Am
Altai Mountains
ghana Valley · XINJIANG · Gobi Desert · EASTERN STEPPE
Pamir · Yanghai · Shirenzigou and Xigou · Ordos loop
Mountains · of Yellow River
Taklamakan Desert
Tarim Basin · Zhaghunluq
Liushui · Guolimu
Wakhan · Hotan · Yellow River
Corridor · Taklamakan Desert
Swat Valley · Tibetan · Xi'an
Plateau
Indus Valley · Gelintang

0   500   1000 mi
0   500   1000   1500 km

6 ORACLE BONES

Grass and grass eaters have always been the key to life on the dry eastern steppe, but those lifeways were fundamentally changed by the arrival of domestic horses. Lying in the rain shadow of the Altai Mountains and marked by frigid temperatures, the Mongolian Plateau presents a punishing environment with many challenges for human life. Nonetheless, the region has been home to people the first modern humans made their way into Asia. During the late Pleistocene, the great dry Gobi was greener and wetter than it is today, and the Mongolian steppe hosted large game like mammoth, camel, horse, aurochs, wooly rhino, and even ostrich.[1] During this time, Mongolia may have played a key role in early *Homo sapiens* movements into East Asia, where modern humans mixed with other preexisting human species.[2] The steppes also hosted wild horses and other equids, including the Przewalski's horse, the wild ass, and the Siberian/East Asian *E. ovodovi*, which despite its characterization as an Ice Age mammal persisted long into the 4th millennium BCE.[3]

As the Ice Age ended and the climate warmed after the Last Glacial Maximum, Mongolia's rich megafauna populations dwindled. While nearby cultures domesticated plants in the early and middle Holocene and crops such as rice and millet took hold across East Asia, Mongolia remained resistant to irrigation agriculture. The steppe stayed loosely populated, and the area's limited Neolithic cultures dabbling in horticulture or agriculture can in some ways be considered extensions of phenomena taking place in nearby areas of China or Siberia than they were true steppe cultures.

HERDING BEFORE HORSES

The balance of life on the eastern steppe changed with the arrival of grass-eating domestic livestock, which helped transform Mongolia's endless seas of dry grass into a predictable, sustainable form of sustenance. The first evidence for domestic animals in Mongolia dates to the Yamnaya/Afanasievo migration ca. 3000 BCE, when migrant pastoralists first introduced domestic sheep, goat, and cattle to the region. In western Mongolia, the Yamnaya/Afanasievo were supplanted in the 3rd millennium BCE by the Chemurchek culture, another pastoral group, this time with apparent genetic and ancestral ties to southern Central Asia.[4] Like the Yamnaya, the Chemurchek primarily raised sheep, goat, and cattle. They stayed close to the mountain margins of the Altai and made little headway in the deserts and dry grasslands to the east. Curiously, neither of these early herding groups seems to have made a lasting contribution to the gene pool of ancient Mongolia.[5] Meanwhile, in northern Mongolia, the Baikal region of southern Siberia, and perhaps in other areas of the eastern steppe, people apparently continued unabated with a hunting and gathering lifestyle well into the 2nd millennium BCE.[6]

Beyond a few isolated equid bones found at Chemurchek ritual sites and a single campsite (thus far unidentified to the species level), there is no indication that cultures of the early 2nd millennium BCE in East Asia had yet acquired the domestic horse.[7] This is not necessarily from a lack of testing. In fact, our own research team spent several years seeking out evidence of early domestic horses in Mongolia. In 2017, we discovered and excavated a habitation site in the Altai Mountains of western Mongolia's Bayan-Ulgii Province dating to ca. 1650 BCE, one of the first sites of its kind. The site, called Tsagaan Asga, consists of several rectilinear habitation structures and corral-like features filled with fragments of animal bones, ceramics, and bronze or copper slag.[8] Most bones from the site were too fragmented to identify through archaeozoologists' usual technique, which involves making careful morphological comparison with reference specimens. However, we were able to identify many of the animals present through a technique known as collagen fingerprinting, which uses a laser to identify differences in the structure of collagen inside the bone that help distinguish between taxa. Through this analysis, we learned that the inhabitants of Tsagaan Asga raised and ate sheep (*Ovis*) and cattle (*Bos*). At another early site, known as Biluut, we found evidence of sheep and cattle remains burned in a hearth, but no other species were present.[9] Together, the faunal record of Mongolia seems to indicate that horses were missing from the first chapter of the pastoral story in the eastern steppe.

THE FIRST HORSES IN THE MONGOLIAN STEPPE

From the Black Sea steppe, horses and chariots dispersed eastward to the margins of the Mongolian Plateau. In the Minusinsk Basin of south Siberia, adjoining Mongolia to the northwest, horses are first found in burials of the Fedorovo horizon around the 17th century BCE.[10] The Karasuk culture (ca. 1400–1000 BCE) continued the use of domestic horses for transport, as evidenced by the bone bridle cheekpieces that occur regularly at Karasuk sites.[11] Using scientific radiocarbon dating, we directly dated one such Karasuk bridle cheekpiece, made of deer antler, to ca. 1536–1323 cal. BCE. Such artifacts indicate that by the midpoint of the millennium at the very latest, horses and horse transport were practiced in areas of southern Siberia adjoining Mongolia.

As they entered southern Siberia, horses were integrated into a diverse cultural landscape already populated by complex, mixed pastoral economies using both wild and domestic resources. After the Yamnaya incursion, indigenous groups like the Okunev (known for its beautiful, elaborately carved human-like standing stones) also integrated sheep, goat, and cattle into their economy.[12] Throughout the 3rd and into the 2nd millennium BCE, many groups seem to have employed a mix of hunting and herding strategies.[13]

However, our understanding of the human-horse story has changed dramatically in recent years on the basis of new archaeological discoveries. In the summer of 2019, I and a small international team of archaeological scientists began a new project, seeking insights into the lifeways of Mongolia's earliest herding cultures by conducting a survey of high-altitude snow and ice patches in the Altai Mountains. Frozen for thousands of years, permanent ice features in this region are melting at an extraordinary pace due to global climate warming. In the process, these snow and ice features are now kicking out ancient artifacts and animal remains that have been hidden inside for thousands of years. Rare finds like these help document aspects of Mongolia's earliest herding cultures that cannot be found in any other kind of archaeological context.

In the summer of 2022, archaeologists on our team made an amazing discovery. Atop a mountain nearly four thousand meters above sea level and amid dozens of other artifacts including ancient arrows and animal bones, we recovered the remains of ancient horses, including the discarded clippings of a recently trimmed horse hoof, melting from the ice margin. DNA testing showed that earlier finds come from the wild Przewalski's horse. However, the hoof trimming, which dated to around 1350 BCE, must have come from a domestic animal. This hoof was recovered alongside the remains of argali sheep and hunting implements (including intact arrows), suggesting that its presence on the high mountain was probably for the purpose of packing out the remains of the wild argali sheep, which ancient hunters would travel high up the mountain ice to pursue in the summer months. Although earlier specimens may yet be found, this is perhaps the oldest domestic horse ever found in the Mongolian archaeological record. Its presence indicates that domestic horses spread from their point of origin in the Black Sea region all the way to the Mongolian Altai by the 14th century BCE.

THE DEER STONE-KHIRIGSUUR CULTURE

With the arrival of domestic horses, everything changed, and Mongolia's first horse culture exploded onto the scene. Referred to as the Deer Stone-Khirigsuur (DSK) Complex, this culture is named for two related forms of stone monuments that rapidly emerged across a broad region of Mongolia and surrounding territories.[14] The term *deer stone* refers to intricately carved standing stones or stelae that are vaguely anthropomorphic. Ranging from a half meter or so to several meters in height, most deer stones boast a belt with tools and weapons, necklaces, shields, and other accoutrements, which often include tiny horses. Some are even

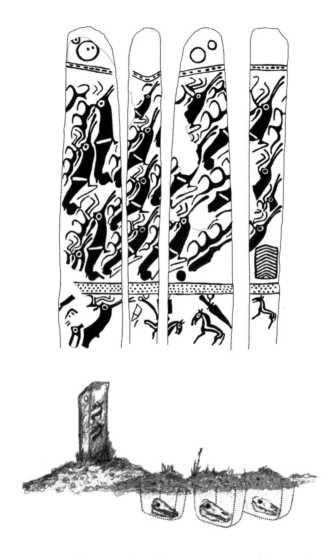

FIGURE 9. A deer stone image from Daagan Del, Zavkhan Province, Mongolia (*top*), showing deer images along with a belt, tools, horse decorations, and anthropomorphic features, including earrings, alongside artist's schematic of three horse burials (*bottom*), facing eastward, buried in small stone mounds along the margins of a deer stone. Image by Dr. J. Bayarsaikhan. Drawing by Barbara Morrison.

carved with human faces, and many are adorned with elaborate stylized flying deer images, from whence they derive their name. While today's deer stone monuments are often worn away through centuries of weathering, research suggests that at least some of the deer images were originally painted with a deep red color.[15] No two deer stones are the same, and some scholars hypothesize that they functioned as memorials for particular ancestors or deceased leaders.

Khirigsuurs, which often co-occur with deer stones at early cemeteries, are stone burial mounds containing human remains. Like Sagsai burials that are sometimes lumped together with them, *khirigsuurs* contain few grave goods and are defined by their central stone mound. In contrast to Sagsai burials, *khirigsuurs* are elaborated with a separate circular or square stone fence-like structure surrounding the mound's periphery. As both deer stones and *khirigsuurs* proliferated widely across the steppes of Mongolia, they reached into southern Russia, eastern Kazakhstan, northwestern China, and even as far as the shores of Lake Issyk-Kul in Kyrgyzstan.[16]

MONGOLIA'S FIRST HORSE CULTURE

Artifacts are scarce at DSK sites, but the archaeozoological record reveals a rich horse-focused pastoral economy and culture, Mongolia's first. At the far margins of both deer stone and *khirigsuur* sites, stone circles contain the burned remains of sheep, goat, and cattle, although horses take center stage.[17] At most deer stones and *khirigsuurs*, stone mounds containing partial skeletons of sacrificed horses were buried around the periphery (usually concentrated at the east or southeast margins of the monument).

Individual horse mounds, which can be up to a few meters in diameter, usually include the head of a horse along with the bones of the neck and either one or both pairs of hooves. In a few instances, additional skeletal elements, like the bones of the tail, suggest that the hide or other parts of the animal were originally included.[18] Some DSK sites have none or only a small handful of horse mounds, while at others, the total number of horse burials range into the hundreds or even thousands.[19]

From studying the skeletons of horses at DSK sites, we know that horses were often killed through a blow to the head using a small pointed axe or other blunt instrument.[20] After receiving its death blow, the horse was sometimes eaten, with meat stripped from the vertebrae, the tongue removed, and many of the elements disarticulated. Parts of the horse that are never found in burials include some of the tastiest and most meat-rich parts of the animal (the torso and lower limbs). These parts of the animal were probably also eaten, perhaps in a ceremonial feast.[21] Small changes to the surface of the horse bones, caused by exposure to weather and sun, suggest that after the animal was killed, it may have been left outside for some period of time before its ultimate burial.[22] Some deer stone or *khirigsuur* monument complexes were clearly constructed in a single event, but careful radiocarbon study suggests that at other sites, horse burials accumulated in discrete events over the span of one two generations.[23]

DSK HORSE TRANSPORT

Because of Mongolia's cold climate, the biomolecules inside bones and teeth are often excellently preserved. Our genomic sequencing of DSK horses and comparison with a large dataset of ancient genomes from across the continent show that DSK horses belong to an early branch of the DOM2 lineage originally found in the Black Sea steppe and are closely related to the first domestic horses from Sintashta.[24] More directly, diagnostic changes to the nasal region of DSK horse skulls show that they experienced prolonged periods of heavy exertion, while pressure on the thin nasal bones during transport also caused remodeling that clearly reveals the use of ancient bridles.[25] Together, these indicators provide conclusive evidence that DSK horses were bridled and used for transportation. Based on patterning in skeletal pathologies, DSK people seem to have preferred adult male horses for transport, although they sometimes also made use of strong mares and animals as young as two years.[26]

HORSES OF THE SUN

Horses and horse transport were the spiritual core of DSK funerary ceremonies and probably the afterlife. While the flying deer images adorning deer stones were intended to carry the spirit to the afterlife, DSK monuments were also clearly structured around the idea of horse transport.[27] At most *khirigsuurs*, a special row of separate horse mounds can be found along the monument's eastern margin. These rows contain remains of only male horses of prime age (between ages five and fifteen years). Compared to other horses found at DSK sites, this special row shows a greater incidence of skeletal pathologies indicative of intensive use in transport and probably represents the favored personal horses of the deceased.[28]

The ancient religious association between horses, chariots, and the sun—perhaps inherited from earlier horse cultures like Sintashta—is also on full display at DSK sites. On some deer stones, four horses are depicted in sequence, galloping toward a sun image. Most horses were buried with their heads facing east and at many *khirigsuurs* are arranged as if pulling the deceased toward the rising sun.[29] The burial plan of some *khirigsuurs*, with their wheel-like corners, call to mind a sort of horse-drawn wheeled tent, similar to the *ger tereg* (yurt carts) used by the great khans and khagans of later steppe empires.[30]

EARLY RIDERS

The importance of domestic horses in DSK culture and the scale of their presence in the archaeological record was absolutely unparalleled in ancient Eurasia. This tremendous burst of horse archaeology led us to wonder whether something new was happening in the ancient world. Did DSK people use chariot transport only, or did these ancient herders also practice an early form of horseback riding?[31] DSK horse burials (see plate 4) lack the skeletal elements most useful for distinguishing transport strategies—the lower vertebrae—which are damaged in different ways by pulling chariots or holding a seated rider. On deer stone carvings themselves, no riders are ever depicted, while at least two deer stones show horses hitched to a two-wheeled chariot.[32] DSK horse skeletons show a strange form of damage to the upper teeth caused by a soft organic mouthpiece whose use during mounted riding might have been especially dangerous.[33] These osteological results suggest that many DSK horses were used to pull chariots.

However, contextual clues from neighboring areas of East and Central Asia suggest that the explosion of DSK horse culture may have been linked to the emergence of mounted horseback riding. The hot sands of the Taklamakan Desert in western China's

Xinjiang Province often faithfully preserve the soft and fragile organic material record of the region's most ancient herding cultures, yielding a wealth of plant material, animal remains, and organic artifacts. Among the amazing material preserved in these Xinjiang desert cemeteries are horse remains from the site of Liushui dating to around 1000 BCE.[34] Unlike earlier Central Asian chariot burials, which usually contain at least a pair of animals, Liushui horses were buried alone.

Excavations at the more northerly cemetery at Yanghai (also dating to the turn of the 1st millennium BCE) have produced the world's oldest trousers.[35] These specialized pants have been reinforced in the crotch area, perhaps for their use in mounted horseback riding.[36] Other tombs contain whips thought to be used as riding goads, as well as balls and sticks that researchers speculate may have been used in equestrian sports.[37] At the site of Zhaghunluq in the Kunlun Mountains, a saddle pad stuffed with white sheep wool was recovered and dated to between 1450 and 900 BCE (although because this date was derived from wood material, it could predate the age of the pad itself).[38]

In 2022, we discovered a petroglyph panel linking the remarkable transformations of DSK horse culture with early reliable mounted horseback riding. At the site of Asgan Khond in western Mongolia's Gobi-Altai Province, we found a Bronze Age carving showing two chariots (one hitched to horses, another lying unhitched). Next to the unhitched chariot, a tiny shield can be seen bearing unique chevron decorations, just like those shown on deer stones, linking this image with the DSK period. Below these two chariots is a single mounted rider holding something like a bow or flag or perhaps pulling a lead connected to the chariot itself. Of course, we cannot rule out the possibility that the rider was added to the composition at a later date. But when considered in context alongside the explosion of DSK horse culture and with the first archaeological evidence for horseback riding like trousers and saddles emerging in Xinjiang only a century or so later, a good argument can be made that this image shows a snapshot of the changing relationship between people and horses at the turn of the 1st millennium BCE.

BRONZE AGE HORSE HERDERS

Ridden horses shifted the balance life in the Mongolian steppes. Horse-mounted DSK herders could move their domestic animals more easily over long distances and in larger numbers, allowing them to make better use of more marginal environments. Taking advantage of their newfound mobility (and perhaps bolstered by a cooler, wetter climate that improved the region's grasslands), DSK herders took herding beyond the mountain margins and out into the dry desert.[39]

With mounted riding, late Bronze Age cultures of the Mongolian steppe were able to raise horses in much greater numbers for meat and milk. In the small handful of garbage piles and middens identified by archaeologists linked to the late Bronze Age, butchered horse bones figure prominently, and evidence for horse-based dairy products also appears in the dietary record gathered from analysis of human dental calculus.[40] From this point onward, fermented mare's milk (also called *koumiss* or, in Mongolian, *airag*) would form a core component of both diet and culture in the eastern steppe.

These first Mongolian horsemen devised new ways of caring for their animals, making tremendous innovations in veterinary care. In 2018, when analyzing a large collection of DSK horses at the National Museum of Mongolia, I and a team of partners uncovered evidence that these Bronze Age herders actually performed rudimentary dental surgery in young horses whose baby or "milk teeth" had come in incorrectly, a condition that would have caused the horses difficulty with feeding and made them difficult to bridle or use in transport.[41] This ranks among the world's oldest evidence for veterinary dentistry and shows that Mongolian herding cultures had begun to use their familiarity with horses to make new leaps forward in horse care, outstripping the veterinary achievements of their contemporaries in the "civilized" agricultural world.

HORSES AND DSK SOCIETY

The introduction of horses in Mongolia does seem to have impacted social inequality, perhaps elevating particular individuals or families to an exalted status. One line of evidence supporting this idea is the size and scale of deer stone and *khirigsuur* complexes in some well-watered valleys of north-central Mongolia, where a single human burial was sometimes accompanied by hundreds of horse mounds representing a tremendous investment of time and labor.[42] DSK carvings hint at a high social value placed on warfare, in contrast with earlier periods. Most stones are decked out with weapons, such as bows and axes, and defensive equipment like shields, and many are decorated with horse images.[43] On the other hand, this does not necessarily mean gender inequity: men and women are found inside *khirigsuur* burials in similar, though not precisely equal, proportions.[44] From what we can tell, then, the shift to mounted horseback riding probably stimulated at least some changes in social organization, giving rise to a new ethos exalting wealth in horses and skill on horseback.

HORSES AND EARLY CHINA

The eruption of horse culture in Mongolia sent tremors across the rest of ancient East Asia. While the horse-centric DSK complex flourished across northern and western Mongolia, horses also spread into nearby cultures occupying eastern Mongolia and the Gobi Desert, appearing in sites of the Ulaanzuukh/Shorgooljin/ Tevsh cultures at essentially the same time as they first emerged at DSK sites.[45] Although Holocene China hosted several wild caballine horse taxa, including *E. przewalskii*, *E. ovodovi*, and *E. lenensis*, the expansion of Mongolia's pastoral cultures was probably the vector that first brought horses to the margins of central China, where the early state of the Shang dynasty ruled supreme.[46] Direct radiocarbon dates on early domestic horses in China is surprisingly rare, but these seem to have arrived in central China sometime after 1300 BCE and most likely after 1150 BCE, trailing the explosive spread of horse culture in the adjoining steppes.[47]

In the late 2nd millennium BCE, archaeological data such as grave goods found in Chinese burials point to increasing connections between Shang China and the steppes and deserts to the north.[48] These connections might have involved movements of people, as northern burial practices (like the unique face-down burial position of some Gobi cultures) spread southward as far as the grassy northern loop of the Yellow River, an area known as the Ordos.[49] With these movements, a near continuous seam of horse-raising cultures now linked the margins of ancient China with the rest of Inner Asia.

Across this margin, cultural exchange flourished. Steppe-linked goods such as animal-headed bronze knives (the same style as those found at Karasuk sites in Russia and carved into the tool-belts of deer stones in Mongolia) began to appear in Chinese tombs alongside other steppe-linked imagery.[50] Political ties also formed; Fu Hao, the favored consort of the Shang emperor Wu Ding, is thought to have originally come from a steppe culture.[51] But the most important good of all to flow across this settled-steppe boundary was the horse, driven by a growing and insatiable demand in China.[52]

ORACLE BONES

Once in China, horses and chariots became popular among the Shang elite, whose aristocratic tombs carved into deep loess deposits have preserved these first horses in stunning detail. From Shang-era tombs, archaeologists have recovered entire chariots and teams of horses in harness. Usually, these horses were sacrificed in teams of two, but sometimes, Shang chariots included up to six animals, along with human charioteers and bronze weaponry. Shang burials also include unique bow-shaped objects thought to be chariot rein hooks, which are identical to images carved along the belt region of many deer stones.[53]

Shang chariots display an unmistakable steppe technological flavor, including spoked wheel construction identical to those depicted in Mongolian and pan-Eurasian petroglyphs.[54] Just like the chariots seen in steppe petroglyphs, each Shang vehicle boasts a single draft pole and a central axle and uses yoke saddles that adapt the yoke to the horse's neck. Despite easy access to bronze, many Shang chariots used an organic bridle mouthpiece, also in line with steppe traditions.[55] Other Shang bits were cast in bronze using a variety of designs that seem to reflect the modification of organic designs to metallurgical production.[56]

Unlike most of Inner Asia during the Bronze Age, Shang China also gives us written records of how these first horses were used. Known as oracle bones, these pictographic inscriptions were typically carved into the shell of a turtle or a cattle scapula and were an early form of writing used for religious divination of important events. After being prepared for divination (sometimes through drilling), the bones would be burned, and the manner in which the bones cracked in the fire was used to read the oracular prediction, which was then sometimes inscribed into the bone. Oracle bone inscriptions often describe agricultural outcomes or difficult social or political decisions facing the king, providing a rich archaeological and historical dataset that also contains clues to the role of horses in Shang society.

Textual references to horses from oracle bones suggest that overall, horses were rare in Shang society, and of recent steppe origin.[57] The oracle-bone script glyph for *chariot* appears ported almost directly from depictions on steppe rock art.[58] The word for horse, rendered as *ma* (马) in modern Chinese, also contains echoes of a steppe past, possibly deriving from a Proto-Indo-Europoean or Proto-Iranic ancestor (as does the Mongolian *mori* [морь/ᠮᠣᠷᠢ]).[59] In oracle bone inscriptions, horses appear to connote great power. The horses known to the Shang were highly diverse in color, with documents mentioning horses of black, piebald, and a reddish or roan color, the latter being especially favored for ritual sacrifice.[60] White horses were also considered particularly auspicious.

HORSES, GRASSLAND, AND SHIFTING GEOPOLITICS IN EAST ASIA

Despite its fondness for horses, the Shang state was not necessarily well positioned to make use of the strategic advantages they offered. While horse riding flourished in the Mongolian steppes to the north, in China only chariots were adopted by the elite. In the agricultural lowlands inhabited by the Shang, obstacles such as disease, poor nutrients, and lack of horse habitat made it difficult to raise large numbers of the animals locally.[61] Those peoples closer to the northern steppes had a tremendous advantage in access to steppe trade and better success raising animals locally. This differential access to horses meant that northern groups were able to amass greater forces of horses and chariots, which created significant power imbalances among Chinese polities. By the middle of the 11th century BCE, people of the western Zhou, thought to have originally come from the steppe, razed the Great City of the Shang with their chariot forces, establishing a new capital near present-day Xi'an in Shaanxi Province.[62] This transition elevated China's northern margins into the driver's seat for political and military power in the region. For centuries to come, the same issue—control over the supply of horses and access to steppe equipment and expertise—would play a guiding role in the political fortunes of early Chinese states (see plate 8).[63]

SUMMARY

In East Asia, the initial spread of the domestic horse was accomplished largely by the herdsman. The explosion of horse culture in the harsh grasslands of the eastern steppe, stimulated by the early adoption of mounted horseback riding, gave birth to a new kind of mobile pastoral lifeway, catalyzing social transformations. Impressive ritual sites and associated horse burials from the region's first horse culture, known as the DSK complex, chronicle the meteoric spread of horses and horse people across the dry margins of the desert and grasslands of East Asia.

The expanding East Asian pastoral sphere brought steppe peoples into contact with the margins of Shang China, where horse exchange helped forge new social, economic, and political connections. Beginning as a relatively rare, elite-only item, chariots and horses quickly became a driving force that rewrote the political landscape, supporting the rise of the western Zhou and toppling the centuries-long Shang rule. As horse riding became increasingly integrated into herding life in the steppes of Mongolia, the adaptation of this new form of transport to combat was poised to reshape social dynamics in the steppe and settled worlds alike.

PLATE 1. Deer stone monuments at Morin Mort in Mongolia's Khangai Mountains (*top*) and associated hoof petroglyphs (*bottom*). Photo by author.

PLATE 2. An early three-toed horse ancestor, *Mesohippus,* which lived thirty to forty million years ago, grazes near a forest margin. Drawing by Judy Peterson.

PLATE 3. Paleolithic cave paintings including horse, bison, and deer at Niaux Grotte in the French Pyrenees, dating as old as seventeen thousand years ago. Photo by author.

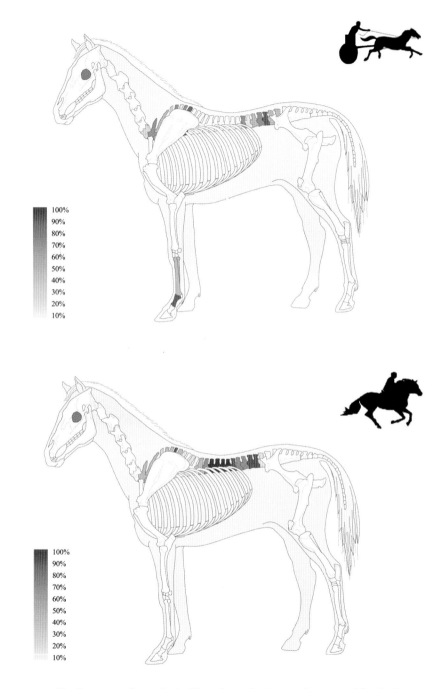

PLATE 4. The frequency of pathological bone formation in horse bones used for chariot transport (*top*) and mounted riding (*bottom*) from ancient Chinese archaeological contexts. Image by Y. Li, modified from original in Zhang et al. 2023.

PLATE 5. Maikop burial showing copper objects once thought to be horse cheekpieces in situ as cattle nose rings, next to a cattle skull. Photo by Anatoliy Kantorovich/Sabine Reinhold.

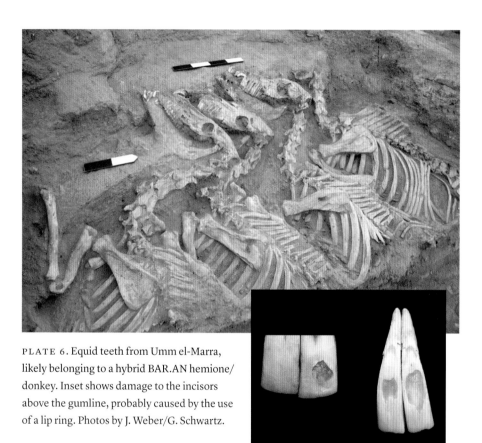

PLATE 6. Equid teeth from Umm el-Marra, likely belonging to a hybrid BAR.AN hemione/ donkey. Inset shows damage to the incisors above the gumline, probably caused by the use of a lip ring. Photos by J. Weber/G. Schwartz.

PLATE 7. Enormous image of Rameses II defeating the Hittites in combat at the battle of Kadesh (13th century BCE) at the Ramesseum, Luxor, and detail showing an early scout/ messenger riding astride. Photo by author.

PLATE 8. Enormous horse and chariot burial pit dating to the 1st millennium BCE, located outside of the modern-day city of Xi'an, China. Photo by author.

PLATE 9. Early proto-saddle of the Pazyryk culture on display at the Hermitage Museum, St. Petersburg, Russia. Photo by author.

PLATE 10. Horse remains from the stable at Casa dei Casti Amanti at Pompeii, frozen in time at the moment of the eruption. Photo by Dr. Johannes Eber.

PLATE 11. Artist's reconstruction of the Greco-Bactrian tapestry pants at Shampula, Xinjiang. Art by Paula López Calle.

PLATE 12. Silver disk with imagery of Greek mythology (Hercules and Omphale), from excavations at Noyon Uul, Mongolia (Erdene-Ochir et al). Photo by G. Galdan.

PLATE 13. Horse petroglyphs adorn a cliffside at Ayrmach-Too near Osh, Kyrgyzstan, located in the Ferghana region, likely the ancestral home of the "heavenly horses." Photo by author.

PLATE 14. A mounted horse messenger rides with his legs supported in cloth loops, depicted on murals from the tombs at Jiayuguan, Gansu, ca. 220–316 CE, in the collections at Gansu Provincial Museum, Lanzhou. Photo by author.

PLATE 15. The frame saddle from Urd Ulaan Uneet, western Mongolia, which may be the world's oldest example. Photo by J. Bayarsaikhan.

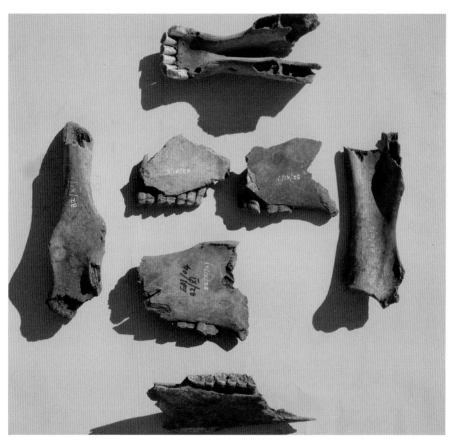

PLATE 16. Archaeological horse remains from the Oyo site of Ede-Ile, in Yorubaland. Photo by Akin Ogundiran.

PLATE 17. Terracotta "Haniwa" horse from the site of Shirafuji Kofun. These Haniwa horses were often arranged around large tombs in Japan during the Kofun period. Photo courtesy the Maebashi City Board of Education.

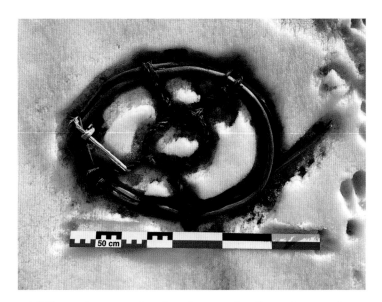

PLATE 18. A Viking-era horse snowshoe from the pass at Lendbreen, Norway, where many ancient artifacts linked with horse transit over the high mountain pass are now emerging from melting ice. Photo courtesy the Espen Finstad/Glacier Archaeology Program.

PLATE 19. Horse bones, teeth, and tack artifacts from the site of Puerto Real, Haiti, in the collections of the Florida Museum of Natural History, one of the earliest archaeological assemblages with domestic horse remains from the Americas. Collections of the Environmental Archaeology Program of the Florida Museum of Natural History, FLMNH–EA Cat. No. 03030231 and 02950514. Photo by author.

PLATE 20. Traffic light during summer festival season in downtown Ulaanbaatar, Mongolia, 2015. Photo by author.

7 HORSEBACK

By the end of the 2nd millennium BCE, domestic horses were integrated into societies across northern Africa, Europe, and Inner Asia. In the steppes, herders were now highly mobile. With speedy and sturdy chariots, people migrated, raided, traveled, and traded across greater distances than ever before. Horses permeated deeply into mythology, religion, and culture. However, horse transport still had many limitations. Chariots were both large and fragile, requiring significant time and energy to build and maintain, and they could bog down easily in the soppy sediments of Europe, Siberia, and South Asia. Outside the steppes, supply of horses could be limited and access difficult, and chariots required at least two animals to operate. Chariots were a luxury of the powerful and wealthy. Inside the steppes, chariots became less common as burial inclusions and receded in archaeological visibility.[1] But just as the impact of the chariot horse appeared to wane, the development of a secure and reliable means of mounted horseback riding in the grasslands and deserts of Inner Asia sparked a second wave of revolutionary changes even more consequential than the first.

The practice of riding a horse was not inherently revolutionary. As we saw in chapter 4, from the early days of donkey domestication in the 3rd millennium BCE, humans had seen the value of climbing astride an equid, a strategy also applied to hemiones or hybrids when the need arose. However, in the early centuries of the horse era following their first domestication in western Asia, riding was mostly undertaken in special situations such as delivering messages or for athletic performance.[2] In a letter to the king of Mari in ancient Syria dated to the first half of the 2nd millennium BCE, a royal advisor urges the regent not to go horse riding but instead to ride a hybrid or a more dignified chariot.[3] Early depictions of mounted horsemanship come from ancient Egypt during the reign of Thutmose III in the 15th century BCE.[4] In Egyptian panels showing a 13th-century chariot battle between the forces of Rameses the Great and the Hittites, a few scattered horsemen can also be seen (see plate 7). These riders, labeled as scouts, apparently rode nude and did not engage in combat. Such finds show that during most of the 2nd millennium BCE, horseback riding was possible (and indeed occasionally practiced) but too dangerous to be reliable in combat or other situations where mobility and nuanced control was needed.

Over the 2nd millennium BCE, incremental technological changes improved the quality of horse control. In tandem with a bridle noseband that assisted in braking, the first Sintashta-style bits with spiked cheekpieces provided far more effective control than the simple nose ring, allowing the driver to compel the animal to turn with a pull of the reins.[5] However, the use of mouthpieces made from soft material would have placed riders in a position of extreme vulnerability were they to wear through or fail.[6] In Mesopotamia and the Near East, the original organic mouthpiece of the horse bridle was adapted into bronze and used to control both horses and donkeys, and similar adaptations also took place in Shang and Zhou China.[7] Metal bar bits enabled new configurations, such as the run-out design using an extension of the mouthpiece to create additional leverage.[8]

One key factor in the development of mounted riding was the selective breeding of domestic horses to make them more tractable and more anatomically stable. Genetic research shows that early domestication entailed strong selective pressures focused on reducing aggression and on strengthening the horse's sensitive spinal area.[9] These selective efforts helped make riding more feasible in high-stress situations.

Among the most important technological developments of the 2nd millennium BCE was the creation of a new configuration of metal bit. The jointed snaffle mouthpiece consists of two metal "canons" joined in the center; when the reins are pulled, the bit exerts pressure on the sensitive corners of the mouth. In experienced hands, this new design can permit more nuanced communication between rider and horse and also make the action of the reins more difficult for the horse to avoid. Dating the first appearance of the jointed metal bit design is difficult, but archaeological discoveries thought to date as early as the 15th–14th centuries BCE are known from Egypt, Turkey, and Armenia.[10] Jointed mouthpieces of a slightly different configuration were also developed, perhaps independently, in China during the Zhou dynasty.[11]

The earliest mounted horseback riding by steppe herders probably took place without special innovations in bridle technology. Although jointed bronze bits appear alongside early riding horses in Xinjiang, damage patterns from DSK horses in Mongolia dating to the late 2nd millennium BCE suggest that these horses were controlled with a simple organic mouthpiece and that chariots were still preferred for combat.[12] If used only in comparatively low-stress situations such as herding and hunting, the first riding mounts of the steppe likely did not require the technological advantages in control provided by jointed metal bits.

EARLY HORSE CAVALRY

By the early 1st millennium BCE, new bridle technologies were widely adopted even deep in the heart of Inner Asia, where early experimentations with riding likely began centuries before. Excavations at the early 1st-millennium BCE archaeological site of Arzhan, in Siberia, revealed several enormous, wheel-shaped royal burials filled with sacrificial horse remains. Horses at Arzhan were also found with jointed metal bits, some of which were paired with boar tusks as cheekpieces or as bridle decorations.[13] Horses from the oldest feature at the site, known as Arzhan 1, were too damaged to be subjected to paleopathological study, but those from the slightly later burial of Arzhan 2 exhibit vertebral damage patterns consistent with their use in mounted horseback riding rather than in pulling chariots.[14]

These advances in control seem to have been instrumental in promoting horseback riding from a specialized activity performed by scouts, messengers, and herders into a pan-Eurasian military approach. The first direct historical depictions of cavalry may be Assyrian murals dating to the 9th century BCE, which show early riding in stunning, high-resolution detail. These carvings depict early cavalry as essentially a chariot team minus the chariot, wherein a team of two horses and two riders worked together in pairs, one controlling the animals while the other engaged in combat.[15] Lacking a true saddle or stirrup, these first riders stayed astride using only a leg grip. The assistance of a teammate and the additional control provided by the metal snaffle bit made control

FIGURE 10. Assyrian relief panel from Nimrud dating to the reign of Ashurnasirpal II showing early cavalry as pairs of riders, one assisting with the reins while the other shoots. Drawing by D. Chechushkova.

FIGURE 11. Jointed metal snaffle bit, an innovation that may have provided increased control over horses and enabled the increased reliance on mounted riding during stressful situations, like combat, in the early 1st millennium BCE. Drawing by D. Chechushkova.

over ridden horses effective enough that for the first time in recorded history, they could be used in battle.

While Assyrian reliefs document the practice first, Assyrians themselves probably adopted the idea of mounted combat from adjoining peoples in northwest Iran and beyond.[16] At the turn of the 1st millennium BCE, a broad swath of Iran, the Caucasus, and the broader Caspian region appear to show an increased archaeological footprint of steppe-linked cultural features, such as a high frequency of horse burials and horse equipment.[17] At Hasanlu, in northwestern Iran, a catastrophic battle event ca. 800 BCE caused the collapse of a horse stable, preserving a snapshot into the use of horses at this time.[18] Ivory carvings from artifacts at this site show horses being used for both chariots and riding, while other finds include elaborate horse bridles and other gear, like a metal chamfron (a battle helmet designed to protect a horse's head). Most probably, this increased footprint of horses and horse culture reflects the expansion of steppe cultures to the margins of western Asia on the heels of a growing movement to use ridden horses in combat.

EARLY EQUESTRIAN CULTURES OF INNER ASIA

During the first few centuries of the 1st millennium BCE, cavalry displaced chariotry as the predominant mode of transport across the continent. As it did so, equestrian peoples and equestrian polities emerged across the steppes and deserts. Inner Asia became increasingly integrated as a coherent culture area across which ideas, technology, and people moved fluidly. From the early 1st millennium BCE, many steppe cultures began to exhibit a diverse mix of pan-Eurasian influences, particularly a kind of animalist art tradition that blended stylistic influences as far-ranging as Mongolia and Persia.

These material confluences during the 1st millennium BCE probably reflected real population movements; recent large-scale sequencing of human genomes across the continent show that this time period saw a new flow of people and genes from the Altai region into the steppes of Kazakhstan and beyond.[19] One example showcasing these colliding cultural influences comes from recent excavations at the site of Tunnug 1, near Arzhan, in Russian Tuva. Scholars describe the burial as a mixture of DSK design components from Mongolia with aspects of influence from Central Asia, such as the use of clay in construction.[20] The spread of mounted riding expanded the geographic horizons for the ancient peoples of the steppe, allowing people, objects, and ideas to circulate over increasingly larger spheres and to converge in new and important ways.

THE EMERGENCE OF STEPPE HORSE CULTURE

The early riders of the steppe began to innovate with horse rearing, honing the horse into an even more nuanced and effective tool in transport and combat. The early DOM2 horse was stocky, strong, and fast, but it lacked important qualities needed for effective riding. Genomic data show that as cavalry replaced chariots, a major uptick in selection took place in Central Asia.[21] Riding horses were bred for musculature, endurance, speed, and even aesthetics. Genomic reconstructions show that during this time, a much wider range of coat colors, including the beautiful yellow palomino color, emerged across the region.[22] Skeletal estimates also show growing cultural preference for taller horses, whose longer strides and taller stature gave mounted warriors an advantage in combat.[23] Archaeological horses from areas such as Kazakhstan, southern Russia, and Xinjiang show that 1st-millennium BCE horses from Inner Asia could reach heights of 1.4 or 1.45 meters, an appreciable boost over early domesticates that may have provided distinct combat advantages to a mounted rider.[24] By around 300 BCE, horses in East Asia had already developed high frequencies of a gene that conferred taller stature.[25]

With ridden horses, steppe herders could raise horses in even greater numbers. At sites in southern Russia, the percentage of bones found in trash deposits rose to 30 percent in the 1st millennium BCE and stayed there well through the Middle Ages.[26] Across Central Asia, a similar pattern developed, with some faunal assemblages rising to include nearly 50 percent horse bones.[27] Horse meat and milk flourished in highly mobile livestock economies.[28] Herders on horseback also gained a tremendous level of flexibility in their living choices, and rock art panels and other data suggest that people likely began to live in both wagon carts and early precursors of the yurt.[29]

When faced with conflict, disease, or ecological degradation, people on horseback could rapidly relocate. A defeated nation of horsemen and horsewomen was never truly vanquished. One impact of the cavalry era was a kind of cultural domino effect wherein consolidation of power in one part of the steppes sent waves of vanquished opponents outward, causing havoc as they moved west.

The first such domino to fall did so in the 8th century BCE, when invading groups known as Cimmerians, probably migrants from the Pontic-Caspian region fleeing pressure from the steppes, conquered a portion of eastern Anatolia and entered the historical record of the Mediterranean world.[30] Westward incursions of steppe cultures would continue throughout the millennium. In one 5th-century BCE trip, Greek historian Herodotus visited the Black Sea, where he encountered a nomadic, livestock-herding people with strange customs and fearsome ability on horseback. These strangers, referred to as the Scythians, became feared and mythologized. In modern parlance, the term has expanded to refer to essentially all early horse-riding peoples of interior Central Asia from the Black Sea to the Altai Mountains.

The Greek world was not the only corner of the continent encountering new mounted threats from the steppe. Horse riding also took root in central and western Europe; from the 8th century BCE, horses appear in sites of the proto-Celt Halstatt and La Tene cultures stretching from Hungary to France. In Slovenia, a number of adult male horse burials at the site of Kobarid include horse-riding equipment and pathological changes to the lower limbs and vertebrae indicative of their use in transport, most likely riding.[31] Human genomic analysis of burials from ancient France suggest a large influx of steppe-associated ancestry in western Europe during the 7th century BCE.[32] As horses spread west, horse riding and horse raising gained a foothold in the political and social landscape of Europe.

Across Central Asia, other early equestrians often lumped together using the name Saka emerged across a broad area stretching from Persia to the Taklamakan.[33] The historical record provides an incomplete and poor chronicle of the Saka, although archaeological excavations of kurgans reveal a rich material culture; Saka nobility clearly enjoyed adorning themselves with animal imagery, gold and bronze objects, and beautiful horse equipment.[34] During the 1st millennium, Central Asian equestrian culture probably also spread to the edges of South Asia, as seen by horse burials in Pakistan's Gandhara Grave culture.[35] In East Asia, horse cavalry also became more visible in the archaeological record. At sites of Mongolia's Slab Burial culture dating to the mid-1st millennium BCE, researchers have recovered other kinds of bridle equipment, including antler cheekpieces and jointed metal bits.[36]

SUMMARY

Like the chariot before, the development of sophisticated mounted horseback riding marked a fundamental shift in the structure of the ancient world. Earlier processes put in motion by the chariot, such as expanding mobility and changes to herding life or social structure, continued or accelerated with mounted cavalry.

The ridden horse continued to accelerate the already deepening social inequality in the steppes, with elite kurgan tombs and lavish grave goods reflecting the emergence of a wealthy elite in many areas during the 1st millennium BCE. However, the social impact of horse riding was also more nuanced. In the steppes, a ridden horse was more accessible than a chariot, requiring a minimum of specialized equipment. When shooting arrows from horseback, a woman was just as dangerous and effective on the battlefield as a man. Based on the high frequency of women found in warrior graves and other historical and archaeological evidence from this period, the early cavalry era seems to have offered many opportunities for women warriors in the Scythian and Saka worlds.[37]

The innovation of mounted riding also brought previously separate areas of the ancient world into regular, intimate, and often unpleasant contact—the first hints of a truly globalized Eurasia. As they permeated the settled margins of the continent, steppe peoples brought with them new language groups and new genes, new ways of living, new ideas, and new technologies. These newfound connections across steppe and desert highways would forge sustained travel routes, diplomatic relationships, and trade between the emerging dynasties of East Asia and the civilizations of the West. And at the heart of these emerging networks sat a new type of society with tremendous influence on the course of human history—the horse nations of Inner Asia.

8 HORSE PEOPLE

The steppe renaissance stimulated by horse domestication and riding set Inner Asia's equestrian cultures on a collision course with settled societies across the continent. These strange new peoples, with their foreign habits, tall pointy hats, and elaborate tattoos, who sometimes lived in mobile dwellings and drank horse milk, were likely a shock to the senses for the denizens of places like ancient Greece, Mesopotamia, or China. Riders from the steppe became the subject of speculation and mythologizing, giving rise to legends like the Amazons, a kingdom of women warriors in ancient Greek mythology likely based on encounters with steppe women in combat.[1] Stereotypes and mischaracterizations of the horse cultures of the Eurasian interior as wild roaming barbarians seeped into Western culture, where they persist into the 21st century.[2]

Archaeological discoveries provide us with mythology-free insights into the daily life of these earliest horsepeople of the steppes, showcasing their complexity and diversity during the 1st millennium BCE. Traces of ancient life ranging from architecture to ancient botanical remains suggest that many of these supposed roaming barbarians were actually accomplished farmers, living relatively sedentary and even settled lives.[3] The diverse peoples and cultures lumped together under the umbrella designations of Scythian and Saka included everything from farmer-herders inhabiting the oases of Central Asia in permanent villages to highly mobile, milk-swigging nomads living in seasonal camps in the frozen Mongolian tundra.

Despite their diversity of cultural practices and lifeways, one stereotype was most certainly true of many Inner Asian peoples in the 1st millennium BCE: they were phenomenal on horseback. A detailed look at horsemanship from this period comes from the archaeological record of the Pazyryk culture, located in the Altai Mountains (spanning areas of Russian Siberia, Kazakhstan, and Mongolia). Here, elite persons were buried in deep tombs that often became permanently frozen, yielding excellent preservation of fragile organic materials.

Finds from Pazyryk tombs give us an unusually rich and detailed snapshot into the life and material culture of Inner Asia in the 1st millennium BCE. Materials such as felt or wooden ornaments normally disintegrate long before archaeologists have the opportunity to recover them, but Pazyryk artifacts include pointed felt caps, pouches, boots, shirts, and blankets made of skin, leather, and wool. Some of the deceased wore elaborate headdresses, false pigtails, or even fake beards. The mummified skin of some Pazyryk individuals shows beautiful tattoos with predator and prey imagery. Wooden bowls, spoons, and ladles were filled with ancient dairy products or provisioned with animal meat for the afterlife. One Pazyryk tomb even included a stringed instrument, a sort of two-stringed harp-style object. The high value placed on mounted combat in Pazyryk culture is evident from the grave goods; most of those interred were buried with the trappings of mounted combat such as daggers, axes, bows, arrows, and even small wooden shields.

Despite their location in the remote and frigid Altai Mountains, Pazyryk grave goods also show another key legacy of horse riding—growing economic and cultural integration with people on opposite sides of the continent. Beautiful tapestries and carpets from burials show stylistic influence or evidence of production in areas as far away as Persia and India, while others show wheeled carts suspected to be of Chinese manufacture.[4] These "barbarians" appear to have enjoyed fineries and global luxury goods on par with any of the ancient civilizations of the 1st millennium BCE.

The most stunning of all the finds at Pazyryk, however, are the horses themselves. While many burials contained only a single horse, wealthier tombs sometimes yielded a dozen or more.[5] Many of the horses interred were outfitted with fine tack and garbed in fantastical costumes. These included gilded bridles decorated with both wild and domestic animals, along with exquisite headdresses that gave the garbed horse the horns of the deer, ibex, or gryphon. Bridle bits were usually made of iron, although bronze was also used.

Beyond the impressive finery, Pazyryk horses reveal some of the sophisticated techniques used in horse care and raising. Some horses were so well preserved that it was clear from remaining soft tissue that they were also castrated.[6] This important practice not only allowed large numbers of breeding-age male horses to be kept in a free-range herd but also made these male horses more docile and easier to control in difficult situations. Hormonal changes linked to castration also probably encouraged these geldings to continue their skeletal development, allowing them to develop slightly longer limbs than an uncastrated stallion, an important advantage in mounted combat. Some Pazyryk horses show osteological features of the pelvis that could also be caused by castration.[7] Tiny healed notches in the ears of Pazyryk horse mummies show that they were often clipped in specific patterns, probably as an indicator of ownership akin to a livestock brand.[8] Our own archaeozoological work shows that in Mongolia, Pazyryk horse herders also performed surgical extractions of the first premolar, or "wolf tooth," that sometimes grows in the interdental space where it can painfully interact with metal bits, a level of veterinary care nearly unheard of in China or western Eurasia at the time.[9] These discoveries show how the steppes and mountains of Inner Asia had become home to wealthy horsemen, making innovations in the care of horses and thriving at the center of an increasingly globalized world.

INNOVATIONS IN HORSE EQUIPMENT

People of the Pazyryk and other horse cultures of the Scythian and Saka spheres were also developing new and better ways to control the horse and fight on horseback. Even as riders in the Mediterranean and Near East rode bareback, steppe peoples began experimenting with an early progenitor of the saddle. From finds at well-preserved cemeteries at Pazyryk and the Taklamakan (in Xinjiang), we can see that these proto-saddles began as simple pads, often stuffed with quilted internal padding made of animal hair. One early find from Yanghai, dated to between 700 and 400 BCE and found in the burial of an adult woman, was made of two lobes of stitched leather stuffed with camel and deer hair, along with straw.[10] Pazyryk saddles further elaborated on the simple pad, sometimes reinforcing the ends with a wooden or bone support and securing the pad to the horse by means of a girth strap around the chest and a crupper, a strap looped around the animal's tail to prevent sliding during downhill travel.[11] Despite their structural simplicity, such proto-saddles were often adorned with elaborate decorations in leather and felt (see plate 9). These improvements in tack helped horsemen stay more secure when mounted.

Armed with effective saddlery, bows and arrows, lightweight shields, and axes, the Scythian and Saka cavalry men and women of the 1st millennium BCE were both agile and deadly. Some horse riders also adopted lightweight forms of scale armor, which protected them while still remaining optimized for both mobility and long hours in the saddle. From a hypothesized origin in southwest Asia, this armor style spread eastward during the early 1st millennium BCE as far as Xinjiang.[12]

IMPACT OF CAVALRY IN WESTERN ASIA

With horses and horsemanship now more central to political power, those with control over productive horse habitat in Inner Asia rose to geopolitical prominence. Conquests by the empires of ancient Greece, Persia, and Rome were buoyed by horses from the grasslands of eastern Europe and Central Asia and sustained by a growing system of horse-based commerce and communication.

With their early adoption of cavalry, the Assyrians conquered most of the area from the Persian Gulf to the Mediterranean, including the kingdom of Egypt and many areas of Anatolia. After the empire's decline in the 7th century BCE, much of the same territory came under control of the Achaemenid Persian Empire, which expanded to cover much of the territory between the Pamir Mountains of Central Asia and the Nile delta by the late 6th century BCE. Persians developed new horse equipment and techniques such as a third rein attached to the bridle noseband that helped produce greater precision when turning.[13] Ridden horses provided crucial infrastructure that helped link the distant pieces of this immense terrestrial empire. Under the leadership of Cyrus, a postal relay system was implemented that connected the imperial capital of Susa in the Persian Gulf region with Anatolia using stations spaced at a distance representing a single day's hard ride on horseback.[14]

Among the most influential military leaders of the 1st millennium BCE was Alexander the Great of Macedon (356–323 BCE). Before beginning his military campaign that would unite Greece with Inner Asia, Alexander first conquered the territory of Thrace, which straddled the westernmost edge of the Eurasian steppe belt, and supplied his growing empire with horses. Alexander's own horse, Bucephalus, may have had ancestry from the Zagros Mountains separating Mesopotamia from Inner Asia.[15]

As heavy cavalry became entrenched in the Mediterranean world, the role of the chariot changed significantly. In *On Horsemanship*, the writer Xenophon describes riders wearing heavy armor, riding bareback, and battling with javelins, a very different undertaking from most mounted combat in the steppes.[16] Lightweight battle chariots, once crucial to warfare in the region during the 2nd millennium BCE, gave way to heavier, tank-like vehicles that could be used in tandem with light cavalry.[17] In Greek and later Roman society, chariots gradually became more important for racing than in travel or battle. A dramatic snapshot into this time of transition was preserved by the eruption of Mount Vesuvius, which buried the Roman villas of Pompeii and Herculaneum in the year 79 CE. The eruption buried a number of stables (see plate 10), and osteological and genomic analysis of the equids within show a range of horses and donkeys both small and large, including three harnessed animals, one adorned with fine red cloth and bronze weights, that met their end while perhaps preparing to pull the large ceremonial chariot discovered in the yard next to it.[18]

IMPACT OF CAVALRY IN EAST ASIA

Eventually, the rising tide of horse culture and cavalry warfare forced the adoption of mounted riding by early Chinese states, too. At the cemeteries of Shirenzigou and Xigou, located at the crossroads of the Taklamakan with the Gobi Desert and the Himalayan Plateau, research by a team of archaeologists and archaeozoologists (including me) demonstrates that people engaged in horseback riding and mounted archery outside the fringes of ancient China by at least the 4th century BCE, if not before. At these sites, diagnostic fractures to the vertebrae of eight horses suggest that these animals were ridden, perhaps with a soft proto-saddle. Interred alongside the dead were quivers of arrows, showing that the deceased were competent archers. Similar finds from the deserts of Xinjiang through the steppes of central Mongolia chronicle the rise of cavalry warfare that began to encroach on the margins of ancient China's early dynasties.

People, ideas, trade, and influence from distant lands began to trickle into the daily life of cultures on the other side of the continent. These emerging links are perhaps best illustrated by a spectacular desert burial at the site of Shampula, near Hotan in western Xinjiang. The burial, which dates to the second half of the 1st millennium BCE, was badly looted, but archaeologists working through the items that were left behind identified a stunning discovery—a pair of trousers apparently fashioned out of a reworked Greek tapestry depicting a centaur and a Greek warrior (see plate 11).[19] How did this remarkable textile with Mediterranean designs reach the tomb of a nomadic horseman on the edge of Taklamakan desert? Through the connectivity of horses and mounted riding, distant corners of Eurasia were suddenly much, much closer than ever before.

For a time, early Chinese states resisted a transition to mounted riding. By the 8th century BCE, the sacking of the Zhou capital by the Quanrong ("dog barbarians") had left a fragmented political landscape characterized by many minor feudal states. In the lower agricultural plains of the Yellow River, unfavorable conditions for local raising of horses meant that access to large numbers of horses was logistically difficult.[20] However, the westernmost Chinese states—Zhao and Qin—bordered on the margins of the Gobi Desert and the eastern steppe and enjoyed access to valuable grassland zones like the Ordos loop of the Yellow River. As raids escalated from horse-raising cultures to north and west, these states were among the first that were forced to grapple with the emerging reality of the cavalry era. Under pressure from steppe raiders, King Wuling of Zhao ordered his army to adopt the dress and tactics of their enemy, demanding that they learn to wear trousers and shoot an arrow from horseback.

Early adoption of cavalry was a boon to western feudal states in China. Although Zhao managed some minor successes on the battlefield, the true winner from the emergence of cavalry in China was the Qin state. Operating from their base of power in Xianyang (near modern day Xi'an), the Qin state leveraged its position at the steppe margin to assemble a large and effective cavalry. Qin warriors were able to expel mounted nomads from the Ordos loop and began a blitz of conquest that would ultimately culminate in the unification of China during the late 3rd century BCE.

The most obvious indicator of the significance of the horse in the unification of China comes from the mausoleum of Emperor Qin Shihuang, the monarch who oversaw the Qin conquests. Following the emperor's death, an enormous mausoleum was constructed containing thousands of terracotta replicas of individual warriors. Among the interred were hundreds of horses, including replica chariot teams and more than 150 full-size cavalry horses.

These terracotta horses, fully equipped with bridle gear, reflect Inner Asian connections. Each horse wears a proto-saddle that could be a near-perfect facsimile of equipment found in Altai and Taklamakan cemeteries, including soft pads, girth straps, and cruppers. The warriors themselves are lightly armored and were probably originally equipped with bows and arrows that would allow them to function as mounted archers.[21] The mausoleum included not only terracotta horses but also real ones; a recent study analyzed twenty-four horse skeletons recovered from an accessory burial pit at the mausoleum showing that tall adult males used for riding were chosen for interment.[22] The absolute numbers of cavalry used by the Qin may have been relatively low compared to the totals of later empires, maybe as few as ten thousand light cavalry and a thousand chariots, but from the mausoleum, it is obvious that even this limited number became a transformative military and cultural force in the unification of early China.[23]

THE FIRST GREAT STEPPE EMPIRE: THE XIONGNU

As horses helped the Qin unite China's disparate states, in the steppes to the north, the rise of cavalry warfare prompted the emergence of a new kind of political integration: the steppe empire. While herders had been joyfully haranguing and raiding Chinese settlements since their earliest days on horseback, their organized expulsion from the Ordos by the Qin may have prompted the nomads of the Mongolian steppes to organize.[24] Under the leadership of Modun Chanyu (perhaps better known to some readers as Shan Yu, the mean-mugging baddie of Disney's animated *Mulan*), a confederacy of pastoral horsemen organized in the heartlands of Mongolia at the turn of the 2nd century BCE.

This group, known as the Xiongnu (Mongolian: *Khünnü*), quickly established authoritative control over the Mongolian steppe and drove other groups, like the Wusun and the Yuezhi, from its margins. But they didn't stop there. The coordinated steppe forces became so devastating that the Qin began a millennia-long effort to build massive, transcontinental walls to keep the Xiongnu and their successor groups out.[25] With their nearly endless supply of horses and horsemen, the Xiongnu also conquered the strategic Gansu corridor and the Tarim Basin of Xinjiang. Meanwhile, with Qin control faltering, China devolved into civil war. In the blink of an eye, a formerly loose confederation of nomadic herders now controlled travel and trade across most of East Asia.

The needs of a bourgeoning empire posed serious logistical challenges in the steppes, where most people lived in distant yurt clusters and gathered only for special events. To manage their emerging empire, the Xiongnu built administrative centers and imperial cities in the heart of the empty steppes. One of the first of these logistical centers is named in Chinese historical records as Longcheng, or Dragon City, known in contemporary Mongolian as Luut. Although lost to the ages for millennia, in 2017, researchers from Mongolia's Ulaanbaatar State University identified the original location of Luut along the banks of the Orkhon River, a lush, centrally located valley nestled in the Khangai Mountains that would serve as the administrative heart of many other steppe empires to follow.[26] In addition to administrative centers, the Xiongnu also built permanent households, temples, and smaller walled compounds. This architectural shift marked a significant departure from the ephemeral structures of the DSK, Slab Burial, Pazyryk, and other earlier cultures in the region.[27]

Just as with Pazyryk, the immense depth of some Xiongnu tombs put them below the line of permafrost, which allowed the contents to be immaculately preserved. Finds at some of these tombs highlight the extraordinary links emerging between the steppe and the rest of Eurasia. At the site of Noyon Uul, excavators recovered blankets with felt applique in the Scytho-Siberian tradition, as well as those with Greco-Bactrian or Hellenistic imagery. One silver plate even shows the Greek hero Hercules making love to the Lydian queen Omphale (see plate 12).[28] At Gol Mod, another Xiongnu royal tomb complex, archaeologists recovered artifacts apparently made of Roman glass and the remains of an entire Chinese elite chariot equipped with a dainty, almost laughably un-Mongolian parasol.[29] Agriculture and domestic grains such as broomcorn millet, wheat, and barley entered the diet of some Xiongnu herders for the first time.[30]

The increasingly cosmopolitan nature of the Xiongnu world can also be seen in changes to its population. Our recent large-scale study of Xiongnu-period DNA shows that this early empire drew in people from Central Asia, Siberia, China, and beyond.[31] In the waning centuries of the 1st millennium BCE, the cold valleys of northern central Mongolia had become deeply intertwined into the social and economic fabric of both sides of the Eurasian continent.

The Xiongnu also continued to innovate in the control and care of horses. Xiongnu horsemen produced high-quality bridles and bits made of bronze and iron while still making use of organic materials such as bone or antler, which could easily serve as a replacement when metal was not available.[32] Excavations at Noyon Uul and other tombs show that Xiongnu riders used even more complex proto-saddles with rigid wooden internal supports that elevated some of the pressure off the vertebral column of the horse, reducing physical stress on both rider and horse and providing greater security during high-stress events like battle.[33]

Earlier livestock-marking systems were replaced by a true brand system wherein a hot iron was used to permanently scar a family-specific symbol, known today as a *tamga,* onto the skin of an animal to indicate ownership.[34] These *tamga* became important indicators of clan identity, particularly among the elite, and began to appear on everything from fancy lacquer vessels to cliffside rock panels during the Xiongnu era. The *tamga* system became a structuring principle of steppe culture that persists into the modern era.

SUMMARY

For better or worse, by the turn of the modern era, horseback riding was a part of life in every corner of Eurasia. With cavalry, the slow trickle of new interactions across the continent that was stimulated by horse domestication and chariots became a torrential flood. Steppe horse cultures drove innovation in horse control and horse equipment and reshaped the steppe horse into a taller, stronger, and more docile animal. Horse warfare stripped authority and geopolitical power from the great agricultural river valleys of the Nile, the Tigris and Euphrates, and the Yellow River and transferred it to those living in the cold tundra, high mountains, dry steppes, and parched deserts. These areas that once served as barriers became valuable zones for raising horses and vibrant centers of trade and political authority. As steppe confederations crystallized into coordinated empires that controlled the emerging transcontinental trade, supply of horses became an imperative for survival.

9 THE SILK AND TEA ROADS

With the might of pastoral powers rising in the steppes, a growing need for horses kicked off a widespread scramble for horses from new regions. Over time, the horse deficit became increasingly dire, particularly in East Asia, where what started as a thriving steppe trade for silks and other fineries became a crisis of geopolitics. In the 2nd century BCE, organized Xiongnu forces reconquered the Ordos loop of the Yellow River and also gained control over Xinjiang and the Gansu corridor, placing most of the quality horse habitat in eastern Eurasia under the budding empire's direct control. Those horses that could be raised in China's core territories were inferior in strength and stamina to their steppe cousins; in 160 BCE, a Chinese official lamented that "in climbing up and down mountains and crossing ravines and mountain torrents, the horse of China cannot compare with those of the [Xiongnu]."[1] Armed horse-mounted raiding from the north could be devastating, and early iterations of the Great Wall built by the Qin and subsequent Han dynasties helped protect against military incursions, but they also limited the movement of livestock.[2] Formal conflict with steppe groups also meant limiting the very trade networks needed to supply Chinese cavalry. To win against the steppes, an army needed horses.

HEAVENLY HORSES

Over the course of the 1st millennium BCE, the mountain zones of the inner continent had become renowned for their tall, strong, and beautiful mounts. With their options dwindling in conflict with the Xiongnu toward the end of the 2nd century BCE, Chinese leadership made a desperate gambit to the west, seeking salvation in the mountains of southern Central Asia. Han Emperor Wudi sent a diplomatic envoy named Zhang Qian far, far westward to what is today Xinjiang, where he hoped to contact the Yuezhi, another pastoral group that had engaged in conflict with the Xiongnu along their western flank and previously served as a major supplier of horses.[3] To the envoy's surprise, however, the Yuezhi had already fled this region on the heels of the Xiongnu assault. Following their trail, Zhang Qian's travels brought him to Ferghana, a fertile valley bracketed by the Tian Shan Mountains and the high Pamir and Alay ranges to the south (see plate 13).

In Ferghana, Zhang Qian found horses that were unusually impressive; they were tall and strong and even appeared to "sweat blood" from their pores when galloping across the deserts and steppes of Inner Asia, giving these legendary animals a celestial sheen (an effect that, perhaps less romantically, might have been prompted by a parasitic infection).[4] When word of these astonishing animals reached China proper, Emperor Wudi sent a series of expeditionary forces to bring back "heavenly horses" as tribute. The first military expedition met with disaster, and while a second, larger force was successful, only fifty heavenly horses survived the return trip to the imperial capital.

Despite the dubious success of these early forays into Central Asia, they provoked throughout China a concerted economic and political interest in western horses. In Chinese artwork, culture, and mythology, the heavenly horses of Central Asia took deep root, a trend perhaps best exemplified by the famous flying horse statuette, a bronze sculpture recovered from a late Han tomb in Gansu and commonly thought to depict Ferghana or Central Asia horses.[5] More formalized trade routes formed along the Tian Shan, across the Pamirs, and even reached through the Wakhan corridor into the Swat and Indus valleys of South Asia.[6] The growing Xiongnu footprint in Central Asia also likely gave them access to these increasingly specialized riding breeds. According to the writings of Chinese historian Sima Qian, writing ca. 100 BCE, the Xiongnu had two types of particularly fine horse, which may have included those imported from Central Asia.[7]

As demand for horses helped drive economic integration from the Chinese side, the cavalry demands of Mediterranean empires in the west like Rome and Carthage also necessitated deeper economic and political connections with horse-rearing zones to the north (Gaul, Iberia, and Thrace), to the south (across northern Africa), and eastward into Asia.[8] From all sides, the economics of horse use helped forge a terrestrial trade system stretching from the British Isles to the Pacific.

INTO THE HIMALAYA

Their cultural and economic value skyrocketing, horses permeated new areas of the ancient world, including Tibet and the high

Himalaya, where horses seem to appear alongside riding in the early or mid-1st millennium BCE.[9] At one site in Gelintang, in western Tibet, a nearly complete horse was buried along with ceramics, metal, and wooden objects dated broadly to ca. 500 BCE–100 CE.[10] One of the oldest archaeological specimens in the region comes from a tomb at Butaxiongqu in Amdo County, northern Tibet. Here, alongside the remains of sheep/goat and domestic dog, archaeologists recovered a horse skull along with lower limbs (a "head-and-hoof" style burial) radiocarbon dated to ca. 760–415 cal. BCE.[11]

The harsh conditions of Himalayan life produced a hardy, strong mount. Modern genomic research suggests that Tibetan horses are closely related to the high-stamina, cold-tolerant horses of the northern steppes, from where they may have originally spread.[12] Horses became entrenched in the culture and economy of highland kingdoms, appearing in sacrificial horse burials along with their equipment across areas like Tibet and Nepal. By the 7th century CE, horses were a key part of funerary traditions, and mounts were sometimes buried in the hundreds in high-status tombs. At Guolimu, elaborate paintings from coffins depict the funerary sacrifice of highly decorated horses at the burial of the deceased for use in the afterlife.[13]

THE TEA HORSE ROAD

Like the heavenly horses of Ferghana, sturdy Himalayan mounts provided a welcome and geographically more accessible alternative to the steppe horses controlled by hostile herders along China's northern frontier. Particularly in times of conflict with steppe groups, demand for horses drove connectivity between China and the highlands of the Himalayas, which began to emerge as an important nexus of horse raising and supply. In the early centuries CE, overland trade routes began to form linking the Tibetan plateau with the lowlands of southern China and with the Ganges River delta on the southern side of the Himalayas. High altitude horse breeders developed a penchant for Chinese goods, particularly tea plants that reached the Tibetan plateau by the early centuries CE.[14] Trade networks linking to the highlands became increasingly important in scope, reaching a fever pitch in the early Middle Ages.[15]

SUMMARY

In the increasingly globalized world that emerged from the 1st millennium BCE, control over horses began to define political and economic fortunes. Equestrian cavalry provided wealth, political power, and military might to emergent powers like the Xiongnu. Facing well-stocked pastoral forces, many agricultural states, particularly in East Asia, were left scrambling to locate enough horsepower to put up a viable defense and to sustain their own economic, military, and social needs. In East and Central Asia, these emerging imbalances drove the first formalization of the Silk Road and the emergence of robust channels of trade and communication between the Tibetan plateau and lowland China.

10 STEPPE EMPIRES

Meanwhile, with horse riding now the predominant instrument of political and economic power in Eurasia, denizens of the steppe marshalled their horsepower into larger, more integrated confederacies spanning vast distances across east and west. The late 1st-millennium rise of the Xiongnu was only the first of many global steppe superpowers that began to shape the world's fortunes around the turn of the Common Era. These vast polities continued to develop new ways of riding, controlling, and caring for horses, while helping create a deeply interconnected continent.

STEPPE INNOVATIONS IN HORSE TECHNOLOGY AND WARFARE

Over and over again, from eastern Europe to Mongolia, new technologies and new strategies in horse riding emerged from horse country. In Europe, people of the Halstatt culture of eastern and central Europe began using the first spurs. These small metal points, attached to the rear of a boot, cajoled horses to high speeds under extreme duress and appear in the archaeological record as early as the 5th century BCE.[1] Later, during the rise of the

Roman Empire in the 1st century BCE, people living in the Balkans, on the margins of the eastern European steppe, developed a new style of bit, now known as the curb bit, using mechanical leverage to generate immense forces on the animal's mouth and lower jaw. Both technologies would later be integrated into Roman cavalry and disseminated across Europe, where they remain in use across the Western world today.[2] However, these examples demonstrate the importance of steppe regions in driving technological progress.

Perhaps the most transformative technologies to emerge from the steppe were the paired innovations of the stirrup and the frame saddle. Today, these items are so synonymous with horseback riding that it is difficult to conceive of horsemanship without them. But during the early centuries of the riding era, neither was in existence.

As cavalry warfare became standard, the challenge of staying mounted in combat was tackled in creative ways. Early iconographic depictions from many contexts in Asia, including painted bricks depicted here from China's tumultuous Six Dynasties period (ca. 220–420 CE), show riders using long fabric loops that probably provided a measure of stability and support to a rider's feet (see plate 14). Other artifacts from Central and South Asia hint at the existence of even more elaborate systems, such as dangling hooks or toe loops, intended to support a rider's feet.[3] While probably increasing the stability of a rider to some degree, it is doubtful that these early foot supports could either hold a standing rider or sustain a rider through a heavy impact.

For those who used them, soft pad saddles were reinforced in various ways, including by sewing pads into unique shapes or securing them with rigid external plates. By the turn of the 1st millennium CE, riders of Inner Asian and East Asian cultures were using saddles with large, rigid external plates similar to the pommel and cantle of modern saddles but without an internal frame.[4] These proto-saddles may have had significant limitations on how much of a given rider's weight they could bear. The proto-saddle was ultimately replaced by a saddle combining a jointed wooden frame (or saddle tree) that elevated the rider off the animal's spine with two metal stirrups. Standard archaeological wisdom places the development of the frame saddle in the 6th century CE, when they also appear across the archaeological record of much of Inner Asia.[5]

The early innovation of stirrups is typically thought to have taken place in China and Korea, where the first stirrup-like objects appear in terracotta models and some burials during the 3rd–4th centuries CE.[6] Because they appear alone rather than in pairs and are shown depicted only the animal's left side, these early stirrups may have originally functioned as aids for mounting into the saddle.[7]

Although archaeological data from this period are harder to come by in the mountains and steppes of Mongolia, recent archaeological discoveries from this region have upended our understanding of the history of the stirrup and the saddle, suggesting an important role for the steppe in their first innovation and spread. A rare iron stirrup, recovered from a burial at the site of Khukh Nuur in northeastern Mongolia, appears to date to the 3rd or 4th centuries, which would make it among the world's oldest known stirrup artifacts.[8] And in 2016, in a small hillside cavern in Khovd, western Mongolia, researchers from the National Museum of Mongolia led by J. Bayarsaikhan responded to a report by local police of a looted burial. Here, Bayarsaikhan and his team recovered the mummified remains of a man and his horse, along with a full set of bridle equipment and other accoutrements. Among the many remarkable objects recovered from the cave, which is known as Urd Ulaan Uneet, researchers found a nearly perfectly preserved wooden frame saddle, complete with a pommel, cantle, and saddletree hand carved from birch bark and laced with horse leather (see plate 15). Working with Dr. Bayarsaikhan, I and a large a team of Mongolian and international scientists conducted precision radiocarbon dating of both the human remains and the saddle itself. Our findings place the saddle's creation sometime during the late 4th or early 5th century CE, making it the oldest known example of a frame saddle ever discovered in the archaeological record. Together, these recent discoveries from Mongolia raise the possibility that the steppes played a central role in the first innovation and spread of both the frame saddle and the stirrup during the early centuries of the Common Era.

INNOVATION AND STEPPE EMPIRES

With the saddle and the stirrup, horse-mounted steppe polities found themselves yet again with significant advantages in combat over their neighbors. Stirrups anchored to a frame saddle helped riders brace in their seat for shock combat, allowing riders to both sustain and deliver heavy blows using weapons like lances, swords, and spears.[9] Both innovations also allowed riders to stand in their seat and improved the stability of intermediate gaits.[10] By the 5th century CE, paired wooden stirrups became common in China, Korea, and beyond, developing into sturdier metal equivalents by the 6th century when they disseminated across a much wider region of Eurasia.[11] In the mid-6th century CE, control over both the eastern and western Eurasian steppes was consolidated by the growing power of the Turkic Khaganate (5th–8th centuries CE), displacing the leadership of the preceding Rouran who fled westward. While these "Blue Turks" established a vast territorial empire eventually stretching from the Pontic-Caspian steppes to Korea, the displaced Rouran spread into eastern Europe, where new genomic analysis links them with the invading "Avars" who ravaged the remnants of the Roman Empire and may have introduced stirrups into Europe.[12]

CLIMATE CHANGE AND HORSE POWER

Because the dry grasslands of Inner Asia receive such little moisture, fluctuations in temperature can have a big impact on the availability of water. When temperatures drop, less water evaporates and the landscape stays wetter. More water and more grass mean more livestock, including horses.

In Mongolia, recent paleoclimate research shows that over the last several thousand years, sustained periods with less exposure to sunlight (known as insolation) translated into higher lake levels and wetter grasslands.[13] These long stretches of favorable climate conditions favored social and political integration in the steppes. A major downturn in solar exposure helped facilitate the rise not only of the Xiongnu but also the Turkic Khaganate and the Great Mongol Empire (ca. 1200 CE). This expansion of grasslands and moisture likely gave pastoral cultures stronger footholds across the lower latitudes of Central Asia and helped them raise large, productive herds of horses. During the rule of Genghis and Kublai Khan, even the barren desertscapes of Xinjiang had become wet, productive oases.[14]

Meanwhile, the same climate shifts often had the opposite effect on the agricultural world. Plummeting global temperatures and declining precipitation strained the harvest yields of early Chinese states, while a similar set of challenges may have hastened the fall of the Roman Empire beginning in the third century CE.[15] Climate cooling buoyed the fortunes of horse-mounted herders, helping them expand into desert margins and form larger, more expansive empires, often at the precise moment when fortunes were flagging in agricultural societies living at the edges of Inner Asia.

With each wave of people and animals out of the steppe, new cultural and biological links formed across the continent—sometimes with catastrophic consequences. The world's first true pandemic, the plague of Justinian, struck the remnants of the Roman and the neo-Persian empires with catastrophic disease alongside the rise of the Turkic Khaganate. Recent genomic research demonstrates that this epidemic was caused by *Y. pestis*, a microorganism harbored by steppe marmots that probably spread from Asia.[16] Paleoclimate cycles that favored livestock also favored plague-heavy rodent populations, creating a perfect storm for pandemic events that ravaged western Eurasia from the fall of Rome through the 14th century.[17] With these cyclical catastrophes, this period is often referred to in Western histories as the Dark Ages. For the horse cultures of the steppe, however, there is no doubt that the age was a golden one.

CITIES IN THE STEPPE

Each time a trans-Eurasian empire was formed in the steppes, loosely populated steppes now had to serve as cultural hubs and administrative centers. Capital cities were built to meet the administrative and economic needs of these immense pastoral polities, vanishing almost without a trace when the special circumstances that necessitated their creation waned. Today, most of these ancient cities are more easily seen from the sky than from the ground, and sheep graze freely over their shallow footprints.

While these temporary capitals are not much to look at today, archaeology shows us that they were once the pulsing heartbeat of a continent, brimming with people and with things transported from around the ancient world. The lush green Orkhon valley of central Mongolia was a favored location in the eastern steppe, where the Xiongnu built their capital, Luut, and the Turkic Khaganate built their eastern capital of Otuken, which later became the Uyghur city of Ordu-Baliq, the largest urban environment that ancient Mongolia had ever seen. Inside the walls of these ancient cities, a visitor would have encountered a thriving variety of temples, palaces, open spaces, and trade shops.[18]

Of all the great polities built on horseback in Inner Asia, none can compare in extent or impact to the great Mongol Empire (ca. 1200–1400 CE), which first began to crystallize in the early 13th century under the conquests of Genghis Khan. When Mongols gathered political and territorial supremacy in the eastern steppe and began to dominate across western and eastern territories, the need for an administrative capital grew rapidly. One early center of the budding empire was the winter palace of Avarga, in eastern Mongolia. Recent excavations of this site by a team from Australian National University and the Mongolian Academy of Sciences revealed a relatively modest site with wooden and packed-earth buildings, where residents ate locally available animals like cattle and sheep, supplemented by horse and goat.[19]

As the empire expanded, the profile of the capital changed dramatically. Following the death of Genghis Khan, his son Ögedei established the city of Kharkhorum in the Orkhon valley alongside the ruins of the great cities of the Xiongnu, Turks, and Uyghurs. At the site, archaeologists have recovered the remains of Chinese kilns, smithing workshops, and production areas for glass and bone objects. The animal bone assemblage shows creatures and foods from all over the world, not only local domestic animals like sheep, goat, cattle, and horse but also wild taxa like marmot, deer, and gazelle. The city's Chinese district had higher proportions of pig, and less emphasis on horse meat. The archaeozoological assemblage even included an elephant tusk, which must have been transported hundreds or thousands of miles from southern Asia to reach the city.

Running an imperial administration also required a different sort of recordkeeping from what was used by the average herder. With every rising empire, the civilizations of the steppe seem to have adopted and used a different alphabet. Some groups, such as the Xianbei (ca. 1st–3rd centuries CE) and the Khitan (10th–12th centuries) adopted the Chinese script to transcribe their language, despite using a very different system of grammar and speech. The Turkic Khaganate used a beautiful runic script, possibly borrowed/ modified from writing systems of western Asia. The Mongol Empire used several writing systems, including the classic Mongol script modified from Uyghur script, which was itself also modified from old Sogdian, along with Chinese and a square, abstracted variant known as the *durvuljin* script. The freewheeling relationship between steppe empires and the written word have made historical documents rare and challenging to grasp in depth, expanding the importance of archaeological data in understanding their impact on world history.

A GLOBAL WORLD SYSTEM

Steppe empires stitched together the disparate pieces of the ancient world. What began as sporadic transfers of people and objects matured into full-fledged and highly formalized systems of exchange. During the Mongol Empire, a formalized postal road was established linking Kharkhorum with Central Asia, the Caspian, and even the far-away Volga region, in what is now western Russia.[20] In total, this network traversed more than sixty thousand kilometers and incorporated more than forty-four thousand horses alongside "pack sheep" and even sled dogs.[21] More than 1,400 formalized stations, known as *yam*, spanned important stretches of this road at distances of roughly twenty kilometers. These stations supplied licensed travelers and messengers carrying the *paiza*, an early facsimile of the passport, with food, shelter, and fresh horses.[22] The Mongol administration also issued paper currency and administrative protection over trade, travel, and communication across the inner continent. Although the empire was merciless on the battlefield, Mongol social policy was often meritocratic and tolerant, enabling people from a variety of backgrounds and religions to coexist.[23]

Despite the high price of military conquest and disease events, the rise of steppe empires also made journeys across the interior common, comfortable, and profitable. From a material lens, Mongol influence even extended across the Pacific—in the High Arctic of Siberia, displacements linked to the Mongol expansion brought people and horses into the frigid northern zones of northeast Asia, where bronze objects (probably repurposed horse equipment) were even exchanged across the Bering Strait into North America.[24]

SUMMARY

Driven by advantages and innovation in horse equipment, horses helped situate steppe polities as the first global superpowers. When large-scale climate shifts disadvantaged agricultural societies in Europe and Asia, the same changes often benefitted livestock economies of the steppes. At the high point of these grand empires, cosmopolitan centers popped up in the heart of Inner Asia, supported by complex infrastructure like the postal relay of the Mongol Empire. In the emerging globalized world, the problems of one corner of the continent could quickly become the problems of another, and even the most extreme geographic barriers were not long able to hold the horse at bay.

11 DESERT AND SAVANNA EMPIRES

As steppe empires drew people together from across the ancient world, the hottest and driest zones of Africa and southwest Asia remained a key social and ecological frontier that initially limited the social impact of the horse before desert and savanna cultures adapted horses and horsemanship to fit a uniquely desert lifestyle.

THE DESERT AND THE HORSE

Early on, horses began as a cold-weather animal, evolved to thrive in the high latitudes of the Black Sea steppe. The first domestic horses were hairy, with both a thicker coat and dark coloration that made it difficult to stay cool in warmer climes.[1] Horses are also far less drought tolerant than most desert animals, requiring between five and ten gallons of drinking water per day.

Despite arriving in northern corners of the peninsula as early as the 2nd millennium BCE, it is perhaps no surprise that horses initially had limited impact in the true deserts of Arabia during the early centuries of horse domestication.[2] Historical records describe Arabian conflicts with Assyrian groups during the 1st millennium

BCE wherein only domestic camels were used.[3] Even when some in northern Arabia began riding the late 1st millennium BCE, the southern subcontinent remained mostly horseless, no doubt in part because many of their transport functions were better served in the hot deserts by domestic dromedary (one-humped camels).[4]

In Africa, too, horse transport did not take at first outside the well-watered margins of the Mediterranean and the Nile. After their initial introduction during the Hyksos incursion around the 18th century BCE, domestic horses were important not only in lower Egypt and the New Kingdom power centers along the upper Nile under rulers like Tutankhamun and Rameses the Great but also in the highland upper stretches of the Nile where 1st-millennium BCE rulers of the Napatan kingdoms buried chariot horses with elaborate wedjat-eye bridle decorations.[5] Horses moved across the seaside rim of North Africa, where by the 1st millennium BCE, local breeds were effective in cavalry victories against the Roman Empire.[6] But in the early centuries of the horse era, the animals made little headway into the vast deserts of the Sahara that lay south.

THE DESERT HORSE

Adaptations by horses and herders living in these hot and dry zones began to overcome the limitations of the steppe horse. Selection for physical traits like shorter hair and dark black skin helped protect the animals from the intense sun, while wider nostrils and tail posture might have also improved oxygen intake and kept the animals cool in desert environments.[7] Just as important as physical adaptations to the desert, however, was the horse's integration into multispecies support systems. When traversing the open desert, horses could be supported with dromedaries. Camels could haul water for their thirsty equine companions and handle most packing and riding tasks, while camel milk and meat provided key sources of sustenance for human riders.[8] As horses became essential in combat, their pairing with desert-adapted animals helped mitigate the challenges of long, dangerous transits through the deserts of Arabia and northern Africa.

Whether because of physical or cultural adaptations to desert life, by the 1st century CE, horses were apparently in use in the southern stretches of Arabia and modern-day Yemen.[9] Horses and horse riders began to appear commonly in Arabian frescoes, wall paintings, and even couch furnishings, and by at least the 4th century CE, they were widely used by Bedouin tribes of the desert interior despite the harsh demands of desert life.[10]

THE HORSES OF THE CALIPHATE

In the early years of the 7th century CE, the Prophet Muhammad began to acquire a powerful new religious following in cities of western Arabia that formed into a powerful political force. By the time of his death in 632 CE, most of Arabia had already been united under Islamic leadership. The influence of the caliphate spread across the Fertile Crescent, the Levant, and North Africa, channeling the enthusiasm of the young faith into military and political conquests and economic success. As military forces moved out of Arabia into the African continent, they rode on the backs of their desert-honed mounts.

In addition to their deep familiarity with desert life, Arabian horsemen also adopted recent technological innovations and expertise in horsemanship from their Asian neighbors. After one early caliphate, known as the Rashidun Caliphate, conquered Persia in the mid-7th century, their forces adopted stirrups, which by this time had become a standard part of the Inner Asian toolkit.[11] Islamic horses became wildly popular in Asia, with recent genomic study showing that beginning in the 7th century CE, bloodlines from Persia and the Arabian peninsula spread as far away as Croatia and Mongolia in a growing network of horse exchange.[12]

Though the leadership and politics of the Islamic world fluctuated, its influence continued to grow, and by the 8th century, territory controlled by the Umayyad Caliphate extended across the entirety of northern Africa and into Iberia. Horse raising and trade proliferated across northern Africa, and with the aid of dromedaries, ambitious desert crossings helped bring horses into a rich landscape of west African civilizations for the first time.

THE TRANS-SAHARAN HORSE TRADE

Separated from the Mediterranean by about two thousand miles of desert, the semiarid grasslands known as the Sahel was in fact excellent horse country, and the introduction of domestic horses caused a lightning strike of social transformation on par with any earlier changes in the Eurasian steppes.

The earliest archaeological discoveries of horses south of the Sahara come from the margins of Lake Chad in northern Cameroon and date to the early Islamic period, perhaps as early as the 7th or 8th century CE.[13] Once south of the great desert, horses spread westward to Atlantic west Africa by at least the 10th century CE and eastward to the Horn of Africa by the 11th century. The latter event perhaps coincided with great numbers of people and horses brought to Egypt by Arabian migrants during the Hilal invasion.[14]

HORSE EQUIPMENT

As a result of the initial spread of horses to the Sahel from North Africa, early horse equipment was firmly anchored in the Arabian tradition. Initially, items such as bridles, saddles, and stirrups were probably directly imported as part of the trans-Saharan trade, employing style and technological design that were quintessentially Arabian. The Arabian system included a harsh variant of the curb bit that employed an additional ring to anchor the mouthpiece to the lower jaw even as leverage was forced upward, producing a painful nutcracker-like effect in the horse's mouth. Sahelian cultures also integrated Arabian-inspired spurs to urge horses onward, including both small prick spurs and long-pointed spurs.[15]

Uniquely local horse-riding and equipment traditions also developed in west Africa and the inland Niger delta region, including riding with bitless bridles and horse-body modification instead of saddles that some suspect could predate the adoption of Arabian equipment.[16] In east Africa, horse equipment and styles from India also probably had a meaningful influence.[17]

A mounted Sahelian warrior used a wide range of weaponry, including bows, unique projectiles like throwing knives, and a melee of fighting implements like lances and sabers secured to the saddle by a long strap.[18] While traditions of mounted combat in Eurasia first evolved in the pre-saddle era, the first riders in the Sahel were already deeply familiar with both saddle and stirrup and were therefore well positioned to make use of heavier melee weaponry. It is perhaps not surprising, then, that Sahelian cavalry combat included heavier weaponry like thrusting spears and horse armor.[19]

Those controlling trans-Saharan travel corridors, including the kingdom of Ghana in west Africa, Gao in the middle Niger, and

Kanem along the routes connecting Lake Chad with Tripoli, were the first to reap benefits of horses. The western and central Sudan regions began to support both large trading cities and a growing number of autonomous states with a comparative advantage in access to horses and control over the Saharan trade.

By the early 13th century CE, horse cultures of the western Sudan had become regionally dominant empires. The Mali state, anchored in the band of savanna stretching from the middle Niger to the Atlantic coast, established control over the trade routes to Tunisia and Morocco and the cities of Djenné, Timbuktu, and Gao. Mali established a cavalry force of more than ten thousand horses, using them to control a vast stretch of western Sudan.[20] After the decline of the Mali Empire, the Songhai Empire, anchored in Gao, maintained dominance until the end of the 16th century. In Yorubaland, centered around modern-day Nigeria, dominance on horseback helped the Oyo Empire, which reached its height in the 17th and 18th centuries, with state-sanctioned trading corporations employing professional riders, veterinarians, and cavalry.[21] At the site of Ede-Ile, a colony used by Oyo to maintain control over its southeastern provinces, abundant archaeological deposits of horse bones, including baby horse remains, show that horses were bred locally as part of Oyo maintenance of authority (see plate 16).[22] As in the steppes, the introduction of horses shifted power toward the savanna, elevating those who could sustain larger numbers of horses through horse rearing or trade.

A SECOND BARRIER

Despite thriving in the sub-Saharan grasslands, horses could not continue too much farther south. In hot, humid environments such as those encountered in Africa's rainforest, horses struggle with low fertility and high infant mortality. Pregnant mares and young horses must cope with heat stress and associated health conditions, including viruses, parasite-born disease, or colic, a potentially fatal gastrointestinal issue exacerbated by warmer temperatures.[23] Together, these conditions meant that raising and maintaining horses beyond the woodland savanna was nearly impossible.

An even more severe barrier to the spread of horses into the rainforest was the tsetse fly, a large biting insect widespread in tropical Africa that carries a range of illnesses that are fatal to most domestic livestock. From the very first introduction of Neolithic livestock into Africa during the early Holocene, the severity of insect-borne diseases south of the Sahel served as a formidable barrier to the spread of domestic animals beyond the equator. For horses, an especially nasty cocktail of deadly sicknesses includes dourine, African horse sickness, equine babesiosis, and worst of all, trypanosomiasis.[24] With a fatality rate of up to 100 percent for animals that contract it, this horrific disease results in fever, swollen organs, and eventually death.[25]

Across the expanding trade routes of the Sahara, many products found their way between Mediterranean and sub-Saharan Africa—salt, gold, cowry shells, copper alloys, and countless other goods. Control of this trade was largely in the hands of those with camels and horses. Even though disease made it difficult to raise horses in the woodland savanna and rainforest regions of Africa, this area boasted many large population centers. With a stranglehold on the favorable landscape for raising or trading horses, the cavalry of Africa's grassland cultures presented a fearsome and highly effective force in battle and produced significant inequities in power and control across the region.

With the equipment and innovations of the Arabian horse tradition and control over a robust trans-Saharan trade, horses transformed the kingdoms of western and central Sudan into hotbeds of commerce and centers of authority. Although these processes began with camels centuries before the first horses arrived, emergent empires, buoyed by the supply of domestic horses, leveraged their advantageous economic and ecological positions into political dominance, linking the Sahel into a growing global network system. But although horses flourished in the savanna, formidable disease barriers made it difficult for the civilizations of rainforest Africa to adopt horses in meaningful numbers and kept horses from spreading farther south.

SUMMARY

By the middle of the 2nd millennium CE, domestic horses were deeply embedded in societies from the Pacific to the Atlantic. After transiting the deserts alongside a growing trans-Saharan trade, horses flourished from the Arctic to the southern edges of the Sahel, where wealthy horse empires dominated above a challenging equatorial disease barrier that kept horses from surviving long in central Africa. However, it would take a new technological revolution to bring domestic horses to their global dominance across the final obstacle: the open ocean.

BEAT FOUR

THE WORLD

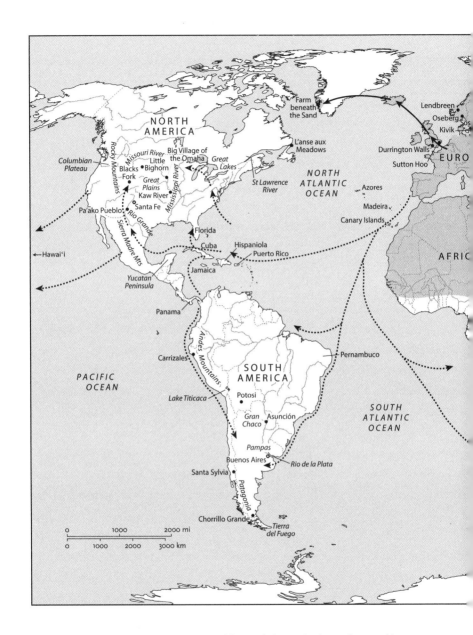

MAP 4. The overseas dispersal of domestic horses in the ancient world. Map by Bill Nelson.

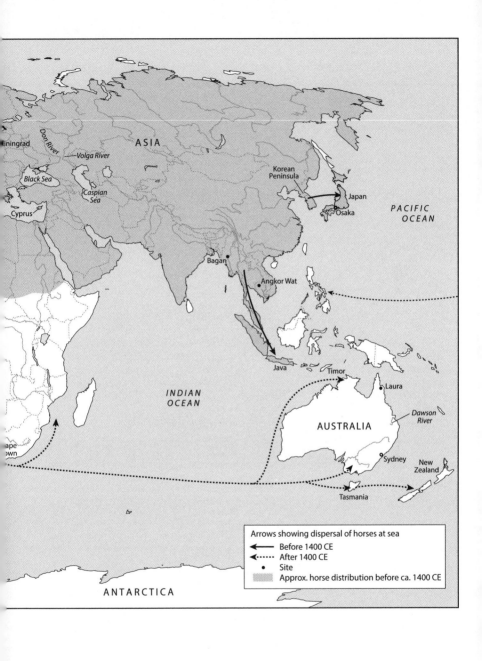

12 OUT TO SEA

With the Islamic expansion bringing them across the Sahara, horses had largely reached the limits of their feasible distribution in Eurasia and North Africa by the 1st millennium CE, but improving technology in open-ocean voyaging would soon bring them to far-flung areas of ancient world.

The maritime transport of horses actually began not long after their initial domestication, as domestic horses were adopted by coastal groups with the capacity for limited ocean travel. When horses spread out of the steppes and into northern Europe during the 2nd millennium BCE, short maritime leaps took the animals across from mainland Europe and into Scandinavia and the British Isles. Archaeological discoveries at Kivik, in southern Sweden, and Durrington Walls, in the United Kingdom, show that protected regions of sea like the southern Baltic and the English Channel were no significant barrier to transporting horses. The latter part of this millennium saw a surge of seafaring activity and boat-based warfare in the Mediterranean. Around 1200 BCE, oceangoing marauders, referred to as the Sea Peoples, probably emanating from Greece and the Aegean, wreaked havoc in Egypt and across the region.[1] Carvings from this period show that Sea Peoples carried with them horses and chariot teams that they used in terrestrial conquests.[2]

With seafaring technology continuing to make strides, oceanic voyages shuttled horses between North Africa and southern Europe, and domestic horses became part of societies in water-bound landmasses like Cyprus.[3] With their advanced shipbuilding, Romans in the late 1st millennium BCE adapted special transport ships with sufficient fodder and water to move horses safely over long sea voyages.[4] By the turn of the Common Era, horses were widely transited across the Mediterranean by merchants and warriors alike.

HORSES AT SEA IN EAST ASIA

In East Asia, the initial spread of horses stopped at the edges of the steppe before reaching the Pacific coast. But alongside the innovation of stirrups and frame saddles and their integration into the eastern steppe, horses and horse equipment became an essential part of life in the Korean Peninsula, too. In the wake of declining Han power in the region in the early 4th century CE, renewed steppe influence brought a wave of horse culture to the peninsula in the form of stone burials, a tradition of goldwork, and elaborate horse trappings.[5] Horses figured prominently in the politics and material culture of the powerful Koguryo state, which came to control parts of Inner Mongolia, Manchuria, and Korea during the 5th century CE. In coastal Korea, people already used composite planked vessels capable of longer-haul voyaging.[6] Soon, coastal networks of trade and exchange brought the first domestic horses from Korea to Japan.

The early oceanic transit of horses in East Asia may have been challenging. In the 4th century, transport vessels would have supported only two rowers side by side and might have been capable of transporting only a single horse at a time.[7] But by the 6th century, advances in shipbuilding had produced larger ships, some with double masts, with the capacity to transport loads of roughly five people and two or three horses at a time along with weapons and food.[8]

Radiocarbon dates on early horses from Japan are rare, but the first mounts might have arrived as early as the 4th century CE. Horse burials are often found in association with large burial mounds that also sometimes contain early saddles and stirrups.[9] One popular funerary offering was the *haniwa* horse, a small terracotta figurine that was placed around large tombs of the Kofun culture, along with other small votives representing elites, subjects, and various animals (see plate 17). In Kofun cosmology, boats and horses seem to have shared a special connection; boat images are often depicted in tandem with coffins and horses on funerary murals, perhaps reflecting their shared role in transit to the afterlife.[10]

Horse rearing became a valuable pursuit in southwestern Japan, particularly in the plains around contemporary Osaka.[11] Stable isotope analysis of horse remains show that as early as the Kofun period, the diet of some Japanese horses was supplemented with domestic grains.[12] While genetic evidence demonstrates the ancestral link between Japanese horses and the steppe, by the late 8th century, Japan had apparently already developed uniquely Japanese mounts with notably long limbs.[13] Over the centuries ahead, horses would retain an important role in Japanese politics, economy, and culture.

THE VIKING AGE

As horse culture expanded in Japan, horses were also spreading across the maritime margins of northern Europe, half a world away. Horses became an integral part of religious beliefs and ceremony. At the cemetery of Aleika 3, in what is today Baltic Russia, horse burials likely date back to the mid-1st millennium BCE and are found in northeast Poland and the Baltic states beginning at ca. 350 BCE.[14] In the Jutland peninsula of Denmark, horses and horse equipment were sacrificed in elaborate lakeshore rituals by the 3rd century CE.[15] These sacrifice events probably involved simulated combat, wherein dozens of wounds were inflicted with various kinds of weapons, including arrows and spears.[16] Across the North Atlantic, it became a common practice to consecrate the construction of a new building by placing the skull of a horse in the wall of the foundation, a tradition that persisted into the modern era.[17]

As the influence of the Roman Empire waned in Europe, invaders from Asia spilled out of the eastern steppe and into the heart of the continent. Catastrophic invasions by the Huns in the 4th century CE and the Avars again in the 7th century sent cultures of eastern Europe cascading westward, beginning an era known as the Migration Period. Germanic populations expanded, and in comparison to the 1st millennium BCE, horses rose dramatically in frequency across western Europe.[18] In Scandinavia, at the site of Sösdala, in southern Sweden, 5th-century CE ritual burials included elaborate gilded horse equipment, probably belonging to an equestrian elite. Along the Baltic coast, horses and horse equipment dominated the elite archaeological record of Lithuania and Latvia.[19] Vandals raided their way into Iberia and northern Africa, seizing Carthage. Goths, Visigoths, and Ostrogoths hassled their unfortunate southern neighbors, while Anglo-Saxons left the northern European continent for the British Isles by sea.

Amid the upheaval, some groups made new strides in ship-building, building larger and more seaworthy vessels. At Mound 1 in Sutton Hoo, in England's Suffolk County, dated to the 7th century CE, an Anglo-Saxon king or elite leader was buried inside a funerary ship roughly eighty feet in length.[20] By the early 9th century, Scandinavian mariners were equipped with masted sail-and-oar-powered vessels.[21] In one superb archaeological example predating 830 CE, archaeologists at the site of Oseberg, in southern Norway, recovered a perfectly preserved wooden ship with a large central mast, indicating that it used a mainsail and wind power.[22]

These larger vessels were sturdy enough to navigate the icy waters of the North Sea, and with the wind in their sails, ambitious Norse sailors set out into the oceans and river systems of northern Europe, raiding and trading as they went. These Vikings settled the coasts of England, Ireland, and the North Atlantic and dominated the Volga and Don River systems of Russia and eastern Europe. Their influence extended into the Black Sea, the Caspian region, and the Mediterranean, where Viking raiders attacked along the coasts of southern Spain, northern Africa, and Italy.[23]

VIKING HORSES

The seafaring supremacy of the Vikings was also paired with equestrian excellence. Genomic comparisons show that Norse horses were closely related to their steppe ancestors and were well adapted to the fierce Scandinavian cold.[24] Archaeological finds on high mountain passes also indicate that Viking horses were used to traverse icy mountain passes, moving both people and agricultural goods with special snowshoes (see plate 18).[25] Viking colonists in the British Isles might even have been among the first to develop a horse capable of ambling (a so-called gaited horse), a unique four-beat gait akin to a slow canter that produces a smooth ride with few vertical disruptions for the rider; some modern Icelandic horses still retain this gait, which in Icelandic is known as the *tölt*.[26] Geneticists are able to identify gaited horses of the past by tracing the presence of a specific allele known as DMRT3, which to date seems to first appear in high frequency among horses in the North Atlantic during the Viking era and may have been strongly selected for by the first settlers in Iceland.[27]

Viking riders made good use of new medieval riding technology, the saddle and the stirrup. Quick hit-and-run archery was not effective in forests, bogs, or fjord lands, but a stirrup-mounted warrior could brace sturdily to deliver fearsome blows at close quarters with swords, maces, axes, spears, and lances.[28] To support heavily laden riders, Norse craftsmen developed sturdier iron stirrups and experimented with new bit and cheekpiece styles.[29] Viking warriors used these tools to their full effect, developing a fearsome style of close-combat cavalry that accompanied their terrestrial raids.

Together, horses and boats were the foundation of Viking warfare. Sometimes, this connection was literal—sails were even prepared with horse grease taken from under the mane.[30] Funerary traditions also reflect this special connection. Horses were occasionally interred in large numbers within ship burials of the Viking elite.[31] At the site of Oseberg, fifteen horses were interred along with the deceased. In one spot, the floorboards of the vessel itself were apparently carved with an image of horses in combat.

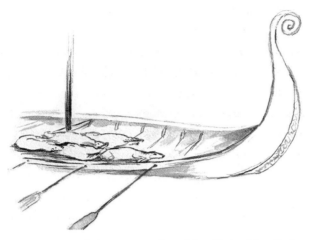

FIGURE 12. Artist's reconstruction of the ship burial at Oseberg, where fifteen horses were laid to rest within the ship along with the deceased. Drawing by Barbara Morrison.

THE FIRST HORSES IN NORTH AMERICA

For the first time, Viking ships could carry horses across the Atlantic. With the climate warming in the waning centuries of the 1st millennium CE, the frigid waters of the North Sea became navigable and attractive for ambitious Norsemen. Isotope study of cremated horse remains from Derbyshire, in the UK, show that as early as the 9th century CE, Viking raiders likely brought their Baltic animals with them as they roved the British Isles.[32] Norse settlement began on the island of Iceland in the late 9th century, establishing horses as a permanent fixture on the island. Horses were particularly influential in early Icelandic culture. Although only around 7 percent of Viking burials in Europe included horses, they were included in roughly 40 percent of Norse Icelandic burials, consisting largely of male horses and the occasional mare.[33] Horses were also a key agricultural animal for Icelandic settlers, with consumption of horse meat so widespread in the 10th century CE that negotiations took place in Parliament over the right to maintain the practice during the formal adoption of Christianity.[34] Under the Viking expansion, domestic horses made a brief partial return to North America for the first time since the Pleistocene.

Norse settlers expanded out of Iceland to the north and west, forming two permanent settlements in southwestern Greenland during the 10th century CE. Archaeological Viking assemblages show that Norse settlers brought the full complement of European domestic animals with them, including horses.[35] At one site in western Greenland known as the Farm beneath the Sands, which was occupied sometime between 1000 and 1400 CE, archaeologists discovered a relatively large quantity of archaeological horse remains, including both juvenile and very old animals, a classic signature of a managed herd.[36]

Despite their return across the Atlantic, horses did not reach beyond Greenland to mainland North America. The deepest foray of Vikings into the continent took place during the 11th century CE, when Norse colonists built a permanent settlement on the coast of Newfoundland, now known as L'anse aux Meadows. Recent archaeological investigations at this site have dated the site's construction with remarkable precision to ca. 1021 CE, but the animal bones found at the site do not include horses or other domestic mammals, only sea mammals and fish.[37] Together, our large genomic analysis of archaeological and modern horse remains from across the western part of North America also showed no genetic evidence for a contribution by Viking horses.[38] Most likely, horses never even made the trip.

A changing climate spelled an end to the Viking experiment with North America. By the 14th century, the warm interlude that facilitated transatlantic travel was waning as the globe entered a prolonged cool period known as the Little Ice Age. Ice began choking the sea routes linking Scandinavia to Iceland and Greenland. The most remote colonies, which were highly dependent on commerce, became increasingly isolated, and Viking life in North America came to a bitter end.

SUMMARY

Even as the Viking era waned, horses in seaworthy ships continued to move overseas out of Eurasia in almost every direction. Arabian and Byzantine horse traders moved horses between Asia, Africa, and India. Religious conflicts between the Christian and the Islamic world sent horse galleys filled with the cavalry of caliphates and crusaders in striking numbers across the Mediterranean.[39] European, Asian, and African powers developed greater skill at keeping horses alive across prolonged sea voyages. As maritime power became synonymous with global power, horses were poised to make the journey to distant lands in a rapidly expanding world.

13 THE RETURN

On boats, horses made a dramatic return to the Americas, their ancestral home. In the wake of both climate and cultural crises of the late Middle Ages, including the cooling of the Little Ice Age, the Mongol invasions, and waves of catastrophic plague, sea travel emerged as a path forward for the cultures of western Europe. Among the first to rebound from these catastrophes were the coastal powers of the Iberian peninsula, which expelled Islamic rulers and began greater forays into the open ocean during the 14th century CE. Portuguese and then Spanish traders augmented their vessels with sails capable of navigating the strong winds and currents of the African coast, borrowing Arabian techniques for celestial navigation. In the first years of the 15th century, Castilian invaders seized the Canary Islands off the Moroccan coast, decimating the island's inhabitants and beginning a lucrative cultivation of the Indian crop sugarcane.[1] Portuguese explorers occupied the island of Madeira a decade or two later, and as early as ca. 1432, horses were brought to the archipelago of the Azores.[2] Horses soon functioned in important social roles on these mid-Atlantic islands, as evidenced by their appearance in early placenames, like Pico dos Ginetes (literally Jinete Peak, referring to a type of light cavalry), and in artwork; a ca. 1503 fresco, recently uncovered hidden inside a church wall in the Azorean town of São Sebastião, could be among the oldest surviving depictions of a domestic horse in the Atlantic.[3] The success of these Atlantic colonial endeavors attracted attention back home, enticing explorers and entrepreneurs westward.

INTO AN OLD "NEW" WORLD

Recognizing the commercial potential of an Atlantic sea route and hoping to reach destinations in Asia, the Genoese merchant Cristóbal Colón outfitted a transatlantic expedition in 1492. The expedition left from Cádiz, in southern Spain, traveling by way of the Canary Islands before hitching a ride on the tropical trade winds westward across the open ocean. Colón, otherwise known as Christopher Columbus, first made landfall in the Bahamas before visiting the islands of Cuba and Hispaniola. Building a small settlement on Hispaniola, Colón returned to Spain to secure funding and supplies for a larger return expedition, which he undertook in 1493. This time, Colón brought horses.

On his second voyage, Colón established settlements in modern-day Dominican Republic. The first settlement, La Navidad, was destroyed in conflicts with the Indigenous Taino people, while the second, La Isabela, was abandoned after only a few years. However, archaeological excavations at La Isabela show that horses had already gained a cultural foothold on the island, revealing horse tack among the objects recovered from its ruins.[4]

With no natural large predators and featuring large tracts of open grassland produced by Indigenous burning practices and Spanish deforestation, Hispaniola was ideally suited to domestic horses and other livestock. Within only a few decades after Colón's arrival, observers reported seeing large numbers of wild horses, donkeys, and other domestic livestock roaming the island.[5] From this first foothold, horses became a thriving part of the island landscape of the Caribbean. Domestic horses were introduced to landmasses small and large, and many islands developed significant populations of feral horses.[6] Horses also proliferated in Spanish domestic herds. Archaeological investigations at the site of Puerto Real, along Haiti's northern coast, first settled in 1503, revealed a number of domestic horse remains amid a huge faunal assemblage (see plate 19). Genetic analysis of horses from the site show connections between early Spanish domestic livestock in Haiti and others across the Atlantic, including the famous Chincoteague ponies along the southeastern coast of North America.[7]

From the Caribbean, domestic horses were carted across the Americas, initially as part of Spanish colonial movements and infrastructure. The expedition to Panama led by Vasco Núñez de Balboa in 1513 alerted Spanish colonizers to a Pacific Ocean link via the isthmus of Panama, prompting permanent settlement and the development of the trans-isthmus Camino Real and Camino de Cruces road systems. Horses and mules were crucial in early transit of people and trade goods across this continental and oceanic divide, and archaeological research at sites linked to this early road system has revealed horse tack such as horseshoes, nails, and spurs.[8] By around 1514, historical records document the presence of horses at the colony of Santa Maria la Antigua del Darién in northern Colombia—at least until they were eaten by jaguars.[9]

In Mexico, the conquistador Hernán Cortés landed in the Yucatan peninsula in 1519, bringing sixteen horses with him, and possibly reinforcing his horse cohort with additional animals from Hispaniola and Jamaica.[10] Despite having such limited numbers, the imposing physical presence of horse cavalry probably factored in his military defeat of the Aztec Empire.[11] Spanish settlers also brought horses to early colonies in Nicaragua and Honduras soon after they arrived in Mexico, and explorer Ponce de León brought fifty steeds with him to southwestern Florida in 1521.[12] In 1526, a large Spanish expedition led by Lucas Vásquez de Ayllón left Hispaniola and undertook a failed settlement likely located on the Cape Fear River in North Carolina.[13] These early forays northward did not apparently succeed in reintroducing domestic horses to the American mainland, however; most horses from early Spanish expeditions to Florida and the East Coast more generally likely died from illness or in conflict with Indigenous peoples.[14]

Despite the best efforts of the Spanish to monopolize control, horses did not long remain exclusively in the hands of colonial invaders. Even as Indigenous nations suffered from disease and the direct genocidal actions of European settlers, many rapidly developed their own relationship with horses. In Hispaniola, the Taino leader Enriquillo may have used horses in anti-Spanish rebellion within a few decades of their first arrival on the island.[15] In Panama, archaeological sites connected to the Indigenous Coclé people and dating to the 16th and early 17th century have yielded the bones of both horses and chicken, and one burial site thought to date to the first decade or two of Spanish colonial control even included the teeth and partial postcranial skeleton of a horse.[16]

Horses proliferated on the mainland, and Spanish hegemony over the animals dwindled. At least on paper, colonies in Mexico had tight regulations against ownership of horses by Indigenous people. But Spanish control over domestic livestock was porous and difficult to maintain.[17] Domestic horses multiplied rapidly, reaching a population in the tens of thousands in the 16th century.[18] Increased Spanish activity in northern Mexico brought horses to the Mexican Altiplano by the mid-16th century, where they were raided by local Chichimec peoples, and also flourished in the wild as *cimarrones*.[19] Native peoples of central and northern Mexico became expert horse riders and livestock traders, exacting concessions from Spanish settlers and harrying trade and travel across these regions.[20]

In 1540, a colonial governor from western Mexico known as Coronado made the first formal European foray into the American Southwest and the southern Great Plains of the United States. With a large team of soldiers and horses, the Coronado expedition made its way up the western front of the Sierra Madre into the Rio Grande valley, where after a series of conflicts with local Pueblo cultures, they crossed the Rocky Mountains and ventured into the plains. Of the hundreds of horses accompanying Coronado, only a handful were female, and this expedition seems to have also failed to introduce a viable population of domestic horses.[21]

When and how did horses enter the Indigenous societies of the Southwest and the plains? Traditional narratives based on European historical documents have usually pointed to a relatively late date, suggesting that even after formal settlement of New Mexico by Spanish colonizers at the turn of the 17th century, Native peoples across the American West were still unable to acquire horses in meaningful numbers until about 1680, when the Pueblo Revolt expelled the Spanish and allowed Native people unfettered access to domestic herds.[22] Working with a large team of researchers from around the world, including scholars and elders from Native nations including the Lakota, Comanche, Pawnee, and Pueblo nations, we set out to test this idea by conducting a large-scale archaeozoological and biomolecular study of ancient horse bones across the American West.

Our new archaeological discoveries show that this narrative is wrong. Although some Pueblo folks lived directly under Spanish control, many others lived in areas with only occasional visits, known as *visitas*, rather than permanent Spanish presence.[23] The site of Paa'ko, on the eastern margin of the Sandia Mountains, was probably one such area, the early Spanish *visita* of San Pedro. Excavations at this site produced a rich faunal assemblage, showing that alongside eating Indigenous dietary staples like domestic turkey, wild rabbit, deer, bison, and pronghorn, the residents of Paa'ko raised and ate sheep, goat, and horses.[24] Importantly, our analysis of horse remains from Paa'ko dated them to the earliest years of the 17th century, suggesting that already by this time, horses were being used in Puebloan communities in northern New Mexico with limited or no colonial supervision. The archaeology is also supported by a more careful read of the historic record; some historical records dating back to the 16th century show the theft of Spanish livestock, including horses, by Indigenous raiders.

Our new data from archaeological science shows that from the Southwest, horses spread quickly north. Despite the fact that historical accounts written by European travelers do not confirm the presence of horses in Native societies in many areas of the American West until the 18th century or beyond, well-dated archaeological horse specimens from Idaho to Wyoming and Kansas now prove that horses were widespread across the plains and the Rocky Mountains by at least the first half of the 17th century.[25]

At sites in Idaho and southern Wyoming, horses linked with Shoshonean and Comanche cultures suggest that horses were exchanged rapidly northward to the Rocky Mountains. At Blacks Fork, a young horse with evidence of tethering and veterinary care was buried in a ceremonial feature along with three coyotes, while at the Kaw River in northeastern Kansas, we identified a pre–Pueblo Revolt horse that had been bridled and ridden. Most strikingly, isotope data suggest that the Kaw horse was fed corn—an Indigenous domestic crop—to get through the winter months.

Our DNA comparisons show that while these earliest Native horses from the Great Plains are of Iberian ancestry, they were ridden, cared for, and deeply integrated into Native cultures long before the first Europeans set foot in the American West, validating oral traditions from some of our Indigenous partners, including Comanche and Pawnee scholars.[26]

Meanwhile, formal settlement of the Eastern Seaboard by British and French settlers created growing population centers of European horses in the eastern portion of the continent that also did not stay long in exclusive colonial control. Many initial colonies on the East Coast failed, with horses meeting their dramatic end as a starvation food. At the colony of Jamestown, the first permanent English settlement, colonists during the so-called Starving Time winter of 1609 ate anything they could to survive. A letter from colonist George Percy described the scene: "Having fed upon horses and other beasts as long as they lasted, we were glad to make shift with vermin, as dogs, cats, rats, and mice."[27]

This event also left an archaeological record; our analysis of archaeological animal bones excavated from a well associated with the Starving Time shows that these first horses were burned, boiled, and hacked into tiny fragments. Even horse teeth were chopped into bits to extract whatever nutrition remained inside the pulp cavity. A continual stream of ships helped bolster the colony with both horses and donkeys. Horses joined early French settlements in the Bay of Fundy as early as 1610, and in the second half of the 17th century, the king began sending significant shipments of horses to colonies across New France.[28] Horses helped furnish colonial settlements along the Great Lakes and the Eastern Seaboard and made their way into the shifting gene pool of both colonial and Indigenous horse herds in the American West.[29]

TRANSFORMATIONS

For many Native nations, especially those in the Southwest and on the plains, domestic horses had an immediate and transformative cultural impact. Caring for horses did not necessarily mean abandoning traditional methods of hunting, gathering, or agriculture, but some groups soon made important changes, like adjusting seasonal movements, to adapt to the needs of horses.[30] Even among people who maintained a permanent village location year-round, horses were often driven to summer and winter pastures, led to water at regular intervals, and fed strategic supplements of corn, hay, or bark during a tough winter.[31] People developed detailed systems of horse medicine and care, and horse hides and tissues became an important source of raw material, useful for everything from clothing to playing cards.[32] Horses could also be an important, if occasional, food source; at the site of Lubbock Lake, in Texas, horses were extensively butchered and stripped of meat and marrow.[33] Native peoples in the West developed their own unique horse lineages, including those with the famous paint and Appaloosa color patterns. Some nations managed reproduction through strategic breeding of stallions.[34] Indigenous riders developed extraordinary prowess on horseback. Horses were specially trained for racing, and across the western United States, horse races became highly significant social, economic, and ritual events.[35]

PLAINS HORSE EQUIPMENT

Indigenous nations of the plains made key technological inventions that reflected a diverse array of roles for the horse in day-to-day life. Spanish colonists arrived with saddles and stirrups, as well as metal ring bits descended from the Arabian and Islamic tradition that employed heavy jaw leverage for mechanical control. When available, this Spanish-style equipment was often used by Native people; ring bits and other horse equipment have been found from early historic archaeological sites across the Southwest and Great Plains, along with other Spanish-sourced metal items, including horse chainmail armor.[36]

However, while Spanish tack was useful, Indigenous groups also designed and made their own horse equipment. Native riders innovated a new and ingeniously simple bridle system that operated via a rawhide loop over the lower jaw. A long rawhide trailing rope extending from these jaw-loop bridles would allow an unhorsed rider to catch and remount their horse.[37] In other cases, a kind of sling suspended between the bridle and the mane would allow a daring horseman to hang *behind* a moving horse during riding and thereby use the animal's body as a shield during combat.[38]

Native horse cultures in the plains also created saddle equipment, including everything from a basic blanket to a complicated frame saddle. Different designs and different materials were tailored to accommodate specific needs and applications—saddles modified for men and for women riders, for hunting and for hauling, and for speed and for comfort. Before guns became widely available, many Native people also employed organic horse armor, examples of which are beautifully illustrated in rock art and other imagery across the plains from Texas to Alberta.[39] Shields and weaponry were adapted for use on horseback, becoming smaller and lighter, with new shapes and designs.

Horses made it simpler, faster, and easier to move and trade, enabling people to range farther afield. Many Plains cultures already used a unique traction system for dog transport, known now as the travois. With horses, the travois was adapted to a larger size, facilitating movement of larger camps/lodges or goods over long distances.[40] Among the Pawnee, tipi rings from before the introduction of the horse appear smaller than those that postdate the horse era, when pack and riding horses permitted larger lodges to be more efficiently stowed and transported.[41] Horses may have also influenced social dynamics. For some cultures, horses became a part of wedding dowries. Transportation was largely women's work, and horse labor helped ease these physical burdens.[42] Across the plains, social systems began to adapt to the changing realities of the horse era.

HORSES IN RITUAL

The spiritual realm became one of the most important roles of the Plains horse. For some groups, horses were (and still are) elaborately painted for ceremony or battle, while others made beautiful horse headdresses and masks that transformed horses into animals like the elk, pronghorn, or buffalo.[43] Horse materials, like the tail, found important uses in regalia and ceremony. It became common across the plains and the Columbian plateau for a recently departed rider's favorite horse or horses to accompany them in death.[44] Among the Pawnee, horses were sometimes offered in sacrifice as an act of prayer or reflection.[45] For the Comanche, a warrior funeral sometimes involved horse sacrifices numbering in the hundreds, rivalling the largest ceremonies of the Eurasian steppes.[46]

THE BUFFALO

In the plains, one of the biggest impacts of the ridden horse was its role in hunting the American bison, *Bison bison*. One of the only giant herbivores of the Ice Age to survive into the Holocene transition, bison were foundational to plains subsistence dating back to the continent's first inhabitants.[47] Before the horse, hunting bison, which could weigh as much as two thousand pounds per individual and attain speeds of thirty-five miles per hour, was a challenging undertaking for hunters without mounts. Bison hunts on foot therefore required meticulous planning, social coordination, and intimate knowledge of migration patterns to track herd movement and familiarity with topographical features, snowbanks, and other natural and artificial barriers to control herd movement enough for successful kills. With horses, however, bison could be effectively taken even by individual hunters. Lightning-quick "buffalo runner" horses became prized for their ability to ride alongside a bison that would be killed by the rider at close range.[48] On horseback, giant herds could now be followed or chased rather than purely anticipated, and meat could be gathered in such quantities that some cultures specialized almost completely in bison hunting. With the extra control provided by expert riding, an ambitious horseman could even cut out a specific animal whose meat was most desirable, usually a tasty younger female.[49]

The Great Plains are horse country, and with horses proliferating, these regions drew people from every direction. From the northeast came the Cheyenne, the Arapaho, the Crow, and the Sioux. Tribes of the Blackfoot Confederacy moved southward from what today is Canada into the high plains of Montana, as did the Assiniboine and the Cree. Groups such as the Salish, Nez Perce, and Shoshone moved eastward from the Great Basin and the Columbian Plateau. The Comanche and the Kiowa moved into the horse-rich southern plains, and from the east came the Pawnee, Wichita, and other Caddoan peoples, Osage, Ponca, Oto, and Omaha.[50] Many non-horse factors probably influenced the decision to make these long-distance migrations, including changing trade networks and social relationships, increasing colonial pressures in eastern North America, and worsening disease outbreaks. But there is no denying that in the plains, buffalo and horses were plentiful; by the 19th century, hundreds of thousands of wild horses roamed the area.[51] Amid the crisis of the colonial world, horses offered both prosperity and opportunity.

As in the Eurasian steppes, Native networks of horse exchange developed that stitched together most of the continent, from

Mexico to the northern plains and from the Pacific to the Mississippi delta.[52] These emerging transcontinental trade lines similarly moved people, animals, goods, and disease. Horses were traded with British and French sources in the waterways of the St. Lawrence and the Great Lakes and with growing British American colonies on the Eastern Seaboard. Our ancient genomic data show that over the centuries, these processes of exchange resulted in mixing between colonial sources and Native horse breeds to produce a unique and diverse horse population.[53]

In a continent thirsty for horses, those with large herds wielded military and economic power. On horseback, hunters could defend and patrol huge hunting grounds, enforcing larger and more coherent boundaries than before.[54] Some archaeological data suggest that the introduction of horses produced a rise in episodes of conflict, along with greater investment in fortifications in some areas of the plains.[55]

Most importantly, horses sustained Indigenous sovereignty. The sophisticated horsemanship and vast herds of horses held by Plains horse cultures gave people territorial autonomy, while their skills in horse riding, herding, and care provided a robust and sustainable economic base that withstood the heavy colonial shocks. Even as genocide and crippling pandemics such as smallpox swept the Great Plains in the 18th and 19th centuries, Indigenous horse nations enjoyed stunning military successes against European powers, extracting political and economic concessions from Spanish, British/American, and French alike.[56] Horse nations like the Comanche and Lakota established dominance over vast geographic regions, bringing European militaries and governments to heel for decades on the battlefield.[57]

FIGURE 13. Shoshone or ancestral Comanche horse petroglyph from southern Wyoming, likely dating to the early decades after the adoption of horses in the early 17th century. Image by P. Doak.

SUMMARY

The archaeological record thus shows us that in a matter of decades, domestic horses escaped the tight control of Spanish settlers, making a triumphant return to their ancestral home in the Great Plains, where the rapid emergence of strong relationships between people and horses disrupted the trajectory of European colonialism and helped sustain the sovereignty of Indigenous nations deep into the modern era.

14 PAMPAS

In South America, too, what began as a purely colonial enterprise soon delivered domestic animals into the hands of Indigenous cultures inhabiting a landscape with deep ancestral connections to the horse. Enticed by whispers of prosperity, Spanish expeditions set out to explore the Pacific coast of South America, bringing horses with them. After a number of failed forays, one expedition led by Francisco Pizarro reached the great Inka Empire, which in 1532 stretched from Colombia to Chile. The conquistador brought with him a cadre of a few hundred men, only some of whom were supplied with horses.[1] Nonetheless, this force was able to capitalize on internal disarray to capture and execute the Inkan ruler Atahualpa and assert colonial authority. Some Inka leaders retreated into the mountains, while conflict and waves of European disease ravaged the populace.

Although 16th-century Spanish accounts made a point of highlighting the supposed fear felt by Native peoples toward horses, archaeological data trace the growing proliferation of horses and donkeys along the margins of the Spanish colonial world in South America, and in some cases, demonstrate the rapid merger of these animals into the lifeways of Andean civilizations.[2] For instance, near

Potosí, in the Bolivian highlands, equid bones dating to the early colonial period have been recovered from both from a lower-elevation spring site with a strong colonial presence and the high-altitude mining site of Cruz Pampa, likely occupied mostly by Indigenous workers.[3] At the site of Carrizales, along the north coast of Peru, animal bones linked to a *reducción*, or forced resettlement of Indigenous inhabitants, also show the presence of horse or mule bones.[4]

As horses multiplied in the Spanish infrastructure of early South American colonies, exclusive control became impossible to maintain, and in many cases, horses found their way into Indigenous societies along the Andean front. Many Andean groups had centuries of experience raising llama and alpaca, which they used for food, wool, and pack transport along mountain roads that helped link Andean empires in communication and trade.[5] In some areas, archaeological fauna suggest that well into the colonial period, camelids were still more economically important than horses, cattle, and other European domesticates, even in some colonial assemblages.[6] However, existing familiarity with large transport animals also made it especially easy to integrate newly introduced horses, which fit neatly into the lifeways of Indigenous societies in the Andes.[7] Affinity for horses continued to spread through societies and networks beyond direct Spanish control, and by the 17th century, Indigenous cultures from Venezuela in the north to the Andean Pacific in the south were raising and riding horses of their own.[8]

TO THE SOUTH

The entrenchment of control in Peru gave the Spanish a platform to expand inland. As they did so, growing populations of horses made the animals increasingly available in southern latitudes. A snapshot into the changing role of animals in the lifeways of early historic Chile comes from the archaeological site of Santa Sylvia, a fortified building built in central Chile during the late 16th century by the Spanish as an *encomienda,* a forced-labor or communal slavery system wherein Indigenous people were compelled to do agricultural work and adopt European language and customs. Animal and plant remains recovered from Santa Sylvia show that horses and other European domesticates were consumed alongside wild taxa. After a short period of use under Spanish control, the site was abandoned.[9]

By the 1550s, Auracanian peoples of northern Chile had already begun riding horses on the battlefield in open combat against the Spanish.[10] Pressure from Auracanian raiders may even have resulted in the abandonment of occupation at Santa Sylvia.[11] Horse images appear widely in the rock art of the early colonial period across northern and central Chile, and horses entered Auracanian funerary and religious ceremonies.[12] When an important person passed on, their horses and equipment were sometimes incorporated into burials, or hung over the graveside.[13] Even as many Indigenous nations were subdued by the Spanish (and later independent Chilean) government, Auracanian peoples maintained political autonomy until nearly the end of the 19th century.

FROM THE EAST

While the Andes were being colonized from the Pacific, Spanish and other European settlers also began incursions on the Atlantic margin of South America, spreading horses as they went. In the 1530s, Portuguese settlements were founded along the rich coasts of Brazil, while Spanish colonists made landfall at the mouth of the Río de la Plata. As early as 1534, Portuguese settlers may have already begun importing horses to the Brazilian coast, via Madeira and the Canary Islands.[14] By 1549, records show the formal importation of horses into northeastern Brazil through the Azores and the Cape Verde Islands to the colonial government.[15] Once again, it did not take long for these horses to reach Native hands. Indigenous people occupying the Gran Chaco, a vast lowland grassland stretching from Bolivia through western Brazil, Paraguay, and northern Argentina, may have acquired domestic horses before 1600.[16] In a 17th-century letter, the colonial governor of Pernambuco mentioned the use of saddled horses and guns by the Tapuya people in combat with the Portuguese, and the fragmentary historic evidence available suggests that Native groups in the 16th and 17th centuries in the grasslands of Brazil used horses in both subsistence and combat and that horses deeply impacted other aspects of daily life like ritual and cosmology.[17] Almost as soon as it began, trade and communication between Spanish colonies in the Atlantic and Pacific was hampered and harried by Indigenous horsemen.

The return of horses to southern latitudes may have been even faster, thanks to a failed colony at the edges of the Argentinian steppe, also known as the Pampas. In 1536, a Spanish expedition led by Pedro de Mendoza founded the first settlement at Buenos Aires, located on the western shore of the Plata. The colony was poorly supplied, however, and suffered from both starvation and conflict with Indigenous people before being quickly abandoned. As the failed settlers left Buenos Aires for the colony of Asunción located farther upriver, they turned loose their livestock, setting horses and cattle free into the grasslands of Argentina. By the time a permanent Spanish presence returned to colonize the southern Plata (ca. 1580 CE), feral horses were abundant, and many may have already been under Native control.

From their platform at the edges of the Plata, horses dispersed very quickly across the grasslands and mountain foothills of Argentina and into Patagonia. While documents mention the presence of domestic horses in northern Patagonia by at least 1621 CE, historic records would suggest that the animals did not penetrate to southern latitudes until well into the 18th century.[18] However, new archaeological data is changing this story; at a site along the Gallegos River deep in southern Patagonia known as Chorrillo Grande 1, researchers recovered a number of ancient horse bones and other artifacts. Our analysis of these horses shows that they were raised locally. Direct dating of some specimens places horses in southern Patagonia by around 1650 CE, a full century before historic documents would identify horses this far south.[19]

HORSE CULTURES OF THE PAMPAS AND PATAGONIA

With horses and feral cattle moving deeper into the southern continent, the Indigenous peoples of Argentina made use of them, both as hunters and as herders. As with the bison of North America, the adoption of mounted riding made the hunting of large herbivores in the Pampas especially effective. Archaeological data show that after horses were introduced, coastal peoples who had previously relied on a heavy marine diet shifted their emphasis toward hunting large inland animals, particularly large wild camelids known as the guanaco. Hunters who had relied primarily on the bow and arrow largely abandoned this technology, switching to the lance and the bola, a fearsome throwing weapon made from a cord and a pair of weights that was especially deadly at subduing livestock and wild game.[20] Historic records from the very beginning of the 18th century are among the last indicators of bow and arrow use among Aónikenk hunters, who moved on thereafter to near exclusive use of the bola.[21] Preexisting hunting techniques for animals like the large, ostrich-like rhea were adapted to horseback, which allowed these big birds to be captured in greater numbers and increased their importance in archaeological assemblages.[22]

HORSE HERDERS

Feral horse populations were booming in the southern steppes, and Indigenous cultures also made use of horses as a foodstuff. Archaeological sites from Atlantic Coast near Buenos Aires all the way to the foothills of the Andes show evidence of horses being butchered, processed, and eaten by Indigenous peoples.[23] At Chorrillo Grande, archaeological horse bones show traces that sometimes even female horses were chosen for butchery and consumption, perhaps reflecting a well-documented Aónikenk preference for mare's meat.[24]

Despite their abundance on the landscape, horse cultures of the Southern Cone also engaged in careful management and care of their horses. Auguste Guinnard, a French entrepreneur who was held captive for years by a Pampean tribe in the mid-19th century, noted skillful castration of young male horses and specific fattening strategies for those horses slated for the butcher's block. Guinnard also observed that horse slaughter was seasonally limited to the colder months, a strategy that maximizes winter survival and spring fertility.[25] Among Patagonian Indigenous groups, another 19th-century traveler, George Chatworth Musters, recorded that among the Aónikenk, special veterinary practices were developed for horses suffering from ailments such as lameness and saddle sores.[26] In some areas, archaeological data show that people even built corrals and enclosures.[27] Both historical and archaeological sources reveal nuanced pastoral knowledge and care of horses by Indigenous groups east of the high Andes.

HORSES IN THE SPIRITUAL REALM

Both historic records and archaeological data show that horses in Patagonia also merged deeply into religious and ceremonial traditions. Among the Aónikenk, Musters noted that horses were slaughtered for important meals and ceremonies, including birth, death, weddings, puberty, coming of age, and whenever a child was injured.[28] Guinnard described attending a funeral where the departed was led on his favorite horse to a prominent ridgetop, after which the horse (and sometimes horse equipment) was sacrificed to join him in the afterlife; Musters noted that parts of the horse skeleton, especially the head, spine, and tail, were placed on a high mountain as an offering after a new marriage.[29] During the voyage of the HMS *Beagle*, even Charles Darwin reported encountering horse bones left as offerings in a revered, solitary tree in the eastern Pampas desert and later seeing horses at gravesites in southern Patagonia.[30] Archaeological horse remains displaying cut marks and other material indicators suggest that feasting on horse meat was a key part of some rituals, while historic records describe the special preparation of meals made from horse meat, blood, and grease.[31]

HORSE TACK AND HORSE GOODS

Peoples of the Southern Cone innovated new ways to control horses. Indigenous nations from the Gran Chaco through southern Patagonia made use of European-made equipment, but also adapted the Spanish bridle, spur, and stirrups into organic versions and manufactured their own horse tack from materials such as horn, wood, and leather.[32] Some horse cultures made horseshoes of guanaco skin and fashioned lassos, girths, and saddles of guanaco or horsehide with a rigid frame.[33] The Aónikenk made horseshoes out of hide for animals navigating tough terrain, along with multicomponent organic bridles with bits, headstalls, stirrups, and spurs, as well as special straps for training horses to stand during dismounts.[34]

Horses also became a crucial source of raw material. European observers note that horsehides were used to produce a wide range of objects, from bolas to tent skins and vases. Horse ribs and hair were used to make violin-like instruments and horse scapulae were recruited to make both shovels and guitars.[35] Horse fat was used for cosmetics and as stomach medicine, while horsehair was used as cordage or to produce trade goods for export.[36]

Native horse cultures in South America developed a deep knowledge of horse behavior that led to tactical brilliance and battlefield superiority, even over European colonists. In one especially brilliant example from Argentina, enormous herds of wild horses would be driven into settler forces with flaming objects tied to their tails, causing European steeds to panic and flee while their riders were routed by Indigenous cavalry.[37]

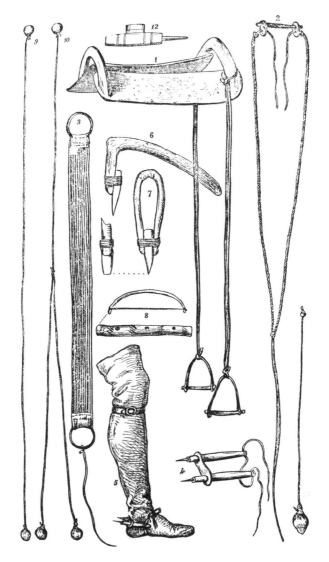

FIGURE 14. Horse equipment and other cultural objects noted by Musters among the Aónikenk in Patagonia, including a saddle with stirrups (1), specialized bridle with organic bit (2), girth straps (3), riding boots with spurs (5), and bolas (9, 10). From Musters, *At Home with the Patagonians.*

SOCIAL TRANSFORMATIONS IN THE
PAMPAS AND PATAGONIA

For many Indigenous groups, horsemanship brought key social changes. Horses became so deeply imbedded in the cultures of the Southern Cone that both horses and riding gear were gifted to infants at birth as their personal possessions.[38] The growing domestic and wild herds of Argentina created animal wealth that drew people from all directions. Aónikenk people moved northward into the plains, and others moved across the Andes from Chile. As in North America, the infusion of horses into preexisting cultural frameworks altered gender dynamics, in this case sometimes placing greater demands on women's workload.[39]

At the same time, groups with horses could use larger dwellings and range over greater distances for travel, trading, or raiding. With their growing wealth on the hoof and their honed skillset in the saddle, Indigenous horse people were empowered to both trade with and raid European settlements. As new colonial constructions were built farther and farther to the south, Indigenous groups began to specialize more in the production of trade items for the European world, such as guanaco hides.[40] Raiding became especially lucrative, generating such material wealth that in the 19th century, it was not uncommon for European settlers to abandon their colonial lives to join the tribes of the Pampas.[41]

SUMMARY

Archaeological data reveal that just as they had on the Great Plains, domestic horses in many areas of South America spread rapidly and were deeply integrated into Indigenous societies long before the first meaningful interaction with European colonists. Upon their arrival, horses became deeply enmeshed in Indigenous lifeways, both as livestock and as a transformative aid to the hunting of large prey like the guanaco and rhea. Horse herds provided raw materials, mobility, and a key reserve of economic and military might and became interwoven into the fabric of ceremony, belief, and culture. Even while colonial settlements, colonial policies, and catastrophic disease disenfranchised Native nations from Panama to the Magellan Strait, Native horsemen regularly routed Spanish and colonial governments on the battlefield and proved to be a key source of sovereignty, wealth, and independence.

15 INTO THE PACIFIC AND DOWN UNDER

Transoceanic voyaging and colonial exploitation across the Atlantic brought prosperity to the Western world and an expanding network of global sea routes carried horses beyond the hard geographic or disease barriers that had once limited their spread. In the southern latitudes of Africa, Australasia, and the Pacific, horses sparked new traditions of horsemanship in nearly every corner of the globe.

EARLY HORSES IN AUSTRALASIA

Before the maritime era, horses had little impact on the Pacific coast of Asia. Although horse rearing became an important regional tradition in upland and mountainous regions like southern China, northern Thailand, and northern Myanmar possibly as early as the 1st millennium BCE, it was only the proliferation of horse trade along the Tea Horse Road that elevated the importance of horses in coastal areas of Southeast Asia.[1] Asian maritime trade networks may have helped horses move across smaller ocean distances and into the southern Pacific, as evidenced by detailed horse relief carvings thought to date to the 8th or 9th century CE in the Buddhist temple of Borobodur, on the Indonesian island of Java.[2] However, because raising horses was more difficult in coastal zones, access to horses may have influenced the northerly and inland location of regional capitals like Bagan in Myanmar (ca. 9th–13th centuries) or Angkor Wat in Cambodia (ca. 12th century CE).[3] During the Middle Ages, meaningful establishment of horse use seems to have eluded the island regions of Southeast Asia.

From the 16th century onward, the increasing presence of European vessels and people—first Portuguese, then Dutch—brought greater numbers of horses along African trading routes. In South Africa, permanent colonial settlements brought horses south of the sub-Saharan disease wall via the coast, and by the 17th century, settlers in the region were importing horses from Southeast Asia, Persia, and Europe in significant numbers.[4]

Despite colonial attempts to restrict and control the possession of horses, they soon escaped into the hands of Indigenous peoples, including the Khoikhoi and the San, as well as creolized settler/Native groups inhabiting the margins of European colonial society who used horses for livestock riding and hunting of wild fauna such as eland.[5] In the mountains of Lesotho, horse rock art imagery shows an important role of horses in men's initiation rituals among the Indigenous Bantu-speaking Basotho.[6] In this region, virtually all adult men traveled on horseback by the early 19th century, and wealth in horses allowed the mountain kingdom to maintain its political independence during European colonization.[7]

Via expanding maritime trade, horses also spread into island zones. Linguistic data point to a dispersal into Madagascar prior to the mid-17th century, and historical records mention horses used by kings in the Indonesian islands of Sulawesi and Sumatra by the early decades of the 17th century CE.[8] By the turn of the 17th century, Spanish traders making transpacific voyages established horses in the Philippines, where the animals quickly proliferated on the islands.[9] In some areas, agricultural deforestation linked to the demands of globalizing trade presented new opportunities for pasturing and grazing for domestic livestock.[10]

THE LAND DOWN UNDER

The Pacific's largest landmass, Australia, was documented in the early 17th century by Dutch explorers, who dubbed the land New Holland but made no formal attempt to settle the continent. Australia had been separated from other terrestrial landmasses for around forty million years. The region's ecological isolation produced a continent filled with weird, wonderful birds and marsupials but few large placental mammals beyond the dingo, whose introduction by ancient Aboriginal groups may have hastened the mainland extinction of the Tasmanian tiger, or thylacine, a doglike marsupial predator.[11]

The first European colonial actions in Australia began under the bourgeoning empire of Great Britain, which began to bolster its presence in the Pacific in the mid-18th century on the heels of nautical victories over other western European powers. Despite the setback of the American Revolution, Britain's Indo-Pacific expansion continued unhindered. British colonies populated largely by exiled convicts were established in wetter coastal areas of the temperate southeast that were most similar to the climate of the British Isles and where European agriculture was more able to thrive, including Sydney, Norfolk Island, and Tasmania during the turn of the 19th century.[12] Along with horses, settlers introduced an array of invasive domestic and wild species, including sheep, goats, cattle, dogs, camels, foxes, and rabbits, kickstarting a cascade of ecological changes that continue today.

When horses first arrived, they were both scarce and in high demand. The journey from Europe was long and perilous, proving fatal for many horses shipped to the early colonies, even those transported from closer ports in the Indo-Pacific. Even if they could be imported alive, horses faced disease and food shortages, while horse owners had to navigate steep restrictions on individual property ownership.

Consequently, the initial growth in Australian horse populations took some decades. The *Derwent Musters* comprise the earliest livestock records from the British colonies in Tasmania, then known as Van Diemen's Land. These lists indicate that the first horse to reach the island was a single mare owned by the military commandant at Risdon Cove in 1803. During the early years of the colony, the military kept tight control over horses. In 1809, the colony boasted only ten horses, six owned by the government and four held by private citizens, two of whom were military officers.

Nonetheless, horse populations grew and were soon bred locally in Tasmania. In 1814, the musters list fifty-two privately held horses, and the colony held its first recreational horse race.[13] By the 1820s, it was finally cheaper to breed a horse locally than to import one. As horse transport became more widely available, newly constructed horse roads linking the towns of colonial Australia became plagued by mounted bushrangers like Ned Kelly, who often raided the remote highways on horseback.

A SECOND INTRODUCTION?

Archaeological data provide some support for the idea that horses could have been introduced to the Australian continent a second time from the northern coast. Off the northern coast of Australia is the island of Timor, where after an initial visit by traders in the 1510s, Portuguese missionaries arrived in the mid-16th century CE, and permanent colonial settlement began in the 17th century, before being overtaken by the Dutch. Direct documentation of the early history of horses in Timor is poor, but archaeological discoveries suggest that horses were around; excavations at the sites of Uai Bobo, Sarasin, and Nikinik 1, which could date as early as the onset of European occupation, have produced horse teeth and bones.[14] More recent work by O'Connor and colleagues also recovered horse bones, including a phalanx exhibiting pathological bone formation dated to ca. 1700 CE at the site of Macapainara.[15] Beginning in the 1820s, a series of British settlements along the northern coast of Australia introduced banteng cattle (*Bos javanicus*) and horses imported from Timor into settlements along the Coburg peninsula. While these settlements failed to gain a solid foothold, the banteng, native to Southeast Asia, were well adapted to the tropical climate and flourished as wild animals. Horses, too, seem to have survived these abandoned settlements, at least for a time. In December 1828, British captain Collet Barker documented the capture of a wild pony near the short-lived settlement of Fort Wellington in Raffles Bay. When an exploratory expedition reached the Dawson River in the northeastern interior in 1844, they reported seeing horses, despite being hundreds of miles from the nearest colony.

ABORIGINAL RELATIONSHIPS WITH
THE DOMESTIC HORSE

The slow initial pace of horse introduction in the southern colonies, coupled with rapid and brutal genocide of Aboriginal peoples, prevented meaningful opportunities for relationships to develop between Indigenous peoples and horses during the early decades of the horse's introduction to Australia's southern colonies. By the 1820s, British settlers in Tasmania, for example, were engaged in open, sustained warfare against the island's Aboriginal inhabitants, and by the early 1830s, most had been killed or forcibly relocated off the island.[16] Nonetheless, horses did make an impact in Tasmanian culture. Tasmanian people developed highly disciplined maneuvers for ambushing, surrounding, and defeating mounted men in combat, and some groups were documented performing ceremonial dances linked to horses.[17] Even under direct colonial subjugation, many Aboriginal peoples developed a close connection with horses. In the 19th century, Aboriginal hunters accompanied Europeans on horseback on expeditions to the interior, and some were conscripted to form cavalry units used in brutal genocidal campaigns against other Indigenous groups.[18] In the continental interior, many agricultural ranching "stations" were staffed and managed by Aboriginal people who became, and remain, expert pastoralists and horsemen.[19]

On mainland Australia, the longer trajectory of colonial subjugation gave greater time for some Aboriginal people to develop independent relationships with horses. Aboriginal cultures often held worldviews that did not correspond to Western ideas of domestic animal management, and during the early decades following the arrival of horses, tack such as saddles and horseshoes often circulated among Indigenous trade networks for their value

FIGURE 15. Artist's rendering of the Giant Horse gallery in Queensland, which includes a horse petroglyph made by an Aboriginal artist, more than six meters in length. Drawing by Barbara Morrison.

as metal tools.[20] However, newspaper records indicate that, as in colonial Queensland in the mid-19th century, Aboriginal people had already succeeded in acquiring and riding horses, using their mounts and self-produced horse equipment such as bark saddle pads to raid and steal cattle.[21]

At northern latitudes, the unsupervised proliferation of Timor horses during the early 19th century may have provided a particularly important, if brief, window of uninterrupted opportunity for cultural connection. Indigenous rock art, particularly from northern areas of the continent, often depicts both horses and riders, sometimes clearly intended to depict Europeans but other times apparently depicting Aboriginal horsemen.[22] Near the small town of Laura, in northern Queensland, one horse panel, known as the "Giant Horse Gallery," is more than six meters in length. These archaeological discoveries speak to the formation of an underappreciated and important cultural link between Aboriginal Australians and horses beyond the boundaries of the British world.

INTO THE PACIFIC

European interests expanded across Australasia in the early 19th century, bringing horses into the open Pacific. British settlers took horses to New Zealand, where they were integrated into Māori societies on the North Island as early as 1814. These first horses functioned as important markers of social prestige in Māori society, eventually becoming a key component of both diplomatic gifting and spirituality.[23] Soon, the Māori were a force to be reckoned with on horseback; the animals were used in armed resistance to British colonial forces in the New Zealand Wars.[24]

While British explorers made their first landfall at Hawai'i decades earlier, it was the former Spanish/Mexican colony of California that first introduced horses to Hawai'i from America in 1803, when a merchant ship brought a stallion and two pregnant mares to Kona and Maui.[25] Also beginning primarily as an elite prestige item, horses took root quickly across the archipelago, as feral populations boomed and Native Hawaiians became accomplished horsemen.[26] By the 1820s, horses were already a significant influence on Native Hawaiian politics and warfare.[27] Some 19th-century European accounts suggested that horses actually outnumbered people and that Native Hawaiians were hardly ever seen on foot rather than horseback.[28] Hawaiian horse equipment blended aspects of the Spanish and Mexican tack tradition with Indigenous additions into its distinct paniolo culture, and horses became integral to sport, agriculture, and transport across the island chain.

SUMMARY

By the end of the 19th century, horses had traversed every mountain range from the Altai to the Andes and had filled every prairie from the Pontic steppe to the Pampas. Domestic equids dotted the streets of coastal ports in every ocean and grazed in grassy meadows on even the most distant landmasses of the Pacific. When British explorers such as Robert Falcon Scott and Ernest Henry Shackleton landed in frigid Antarctica during the first years of the 20th century, horses even set foot (albeit briefly; they were slaughtered and eaten by the struggling explorers) on the icy landmasses near the southern pole.[29] With deep connections between people and horses in every hemisphere, wheels of change were already turning that would bring the era of the horse to a striking end.

16 IRON HORSES

In barely four millennia, horses had gone from a dwindling Ice Age mammal facing down extinction to a thriving domesticate prospering on nearly every large landmass on earth. Horse transport built continent-sized political, cultural, and economic links across empty frigid steppes, dry deserts, and high mountain chains. Horses breathed life into the harsh grasslands of the ancient world, elevating the peoples who inhabited these regions to positions of global influence. Wealth in horses made the agriculturally marginal into military powerhouses, and horses improved the resilience of pastoral lifeways through meat, dairy, and other products. Partnerships between horses and other transport animals and systems, including camels that helped with desert travel and seaworthy vessels that navigated large stretches of open ocean, helped disperse horses across the world. European powers tried to use horses to anchor their colonial exploitation of the Americas, southern Africa, and the Pacific, even while the animals invigorated Indigenous resistance.

As horses reached their greatest global presence, however, a change was afoot. In the 18th and 19th centuries, European societies pivoted dramatically to manufacturing and mechanization. Initially, the burden of this global wave of industrialization was

FIGURE 16. Agricultural horse and plow monument in Worland, Wyoming. Photo by author.

literally laid on the backs of horses. So-called pit ponies began to see widespread use in underground shaft mines across western Europe and its colonies, hauling materials such as coal for long hours in frightful conditions to feed the needs of burgeoning industry. A booming trade in extracting natural resources from the Americas and Australia spurred prospectors to move into the high mountains, where horses labored in mining operations. Agricultural horses were slotted into increasingly complex mechanical systems that turned the gears of the Industrial Revolution, and horse populations boomed around the globe. In Western societies, horses performed harder work with heavier loads.

Innovations in fuel technology and metallurgy and the invention of steam engines also kickstarted mechanization of transport systems. One archaeological anecdote that illustrates the shifting pressures faced by horses during the technological upheaval of the 19th century comes from Lake Champlain in the American Northeast. Here, beginning in the early 19th century, "horseboat" ferries whose turning paddlewheels were powered by an equine treadmill-like contraption moved passengers across the lake. Archaeologists identified the wreck of one such horse ferry at Burlington Bay, Vermont, with excavations revealing horse harness gear, horseshoes, and a treadwheel with heavily worn gears.[1]

In the first half of the 19th century, the first passenger railway was built in Britain, and this new form of locomotion proliferated in Europe and its colonies. The expanded use of these "iron horses" did not immediately lessen the importance of the flesh and blood version. Railroads improved the speed and efficiency of bulk and long-distance transport, but because they traveled a fixed path, for most purposes they had to be connected with other means of short-distance travel. Horses stepped forward to fill these gaps, and specialized breeds of large draft horse, taller and heavier than horses had ever been, gained in popularity across the Western world.

In the United States in the mid-19th century, early railroad sections were unconnected across most of the country's vast interior, which was bridged by horse-drawn mail and stages, including the famous but short-lived Pony Express that linked the Missouri River with California in 1860.[2] Experiments with steam-powered replacements for the horse were amusing but rarely successful. Frank Root, a former messenger on the Overland Stage, related an anecdote in which, amid great pomp and circumstance, an early steam-powered wagon barreled into a historic building before sinking into a muddy stretch of road, never to ride again.[3] Farther from industrialized Europe, horses remained the primary means of transit, agriculture, and other basic tasks much longer.

Horses were still essential in military contexts, too. Despite continuing improvements in industrial weaponry, colonial military forces often found themselves outmatched on the battlefield against Indigenous opponents. In 1876, the US Cavalry faced a massive defeat at the hands of the Lakota, Dakota, Cheyenne, and Arapaho on the Greasy Grass River, today known as the Little Bighorn, in southeastern Montana. When unable to secure victory on horseback, American military detachments opted for genocide, slaughtering entire villages largely comprising elders, women, and children in massacres at Sand Creek, Washita, and elsewhere.

Slowly, mechanized infrastructure turned the tide of colonization against horse country. The improving rail system allowed settlers to be piped wholesale into horse country, where, bolstered by improved supply chains and military protection, they could forcibly convert open grasslands into fenced ranchlands and destroy Indigenous food sources, like the bison, en masse. Growing numbers of settlers meant that Indigenous groups increasingly found themselves living in close proximity to European towns and cities, with greater exposure to European diseases like smallpox and measles that decimated entire communities. To further European control over the Western territories, Native horse herds were slaughtered to quash mobility and resistance. With a dwindling subsistence base and the burden of disease, many Indigenous groups in the United States were forced to accept unfavorable treaties and to settle on reservations where assimilation and repression became a facet of everyday life. Even amid this devastation and the sometimes explicit efforts at eradication, Indigenous horse cultures persisted in the colonial world. In the Great Plains, Native peoples like the Lakota Nation and many others carefully curated traditional horse lineages and traditional knowledge of horsemanship, protecting knowledge and tradition against forceful efforts aimed at their extermination.[4]

FROM HORSES TO HORSEPOWER

In 1885, a German engineer invented the first gasoline-powered motorized car. Except for its lack of horses, the motor car was almost indistinguishable in appearance from a horse carriage. After a brief period as little more than a novelty, the automobile joined the mass wave of mechanization that was stimulated by the onset of the first World War (1914–18). Horses still figured prominently in combat operations during the war itself, with European nations bringing hundreds of thousands of horses and mules to serve in essential lines of communication and troop transport. Horse cavalry even played a central role on the battlefield in decisive cavalry campaigns in the eastern Mediterranean theater.[5] However, postwar production sent automotive production soaring, and soon, commercial motorized vehicles were easily available in Europe, Australia, and the Americas, unseating the horse as the primary means of transport in the Western world.

One of the first places that motor vehicles replaced domestic horses was in urban environments. In the city, raising, feeding, and caring for a horse was an expensive and tremendously messy hassle. Estimates suggest that in 1880, horses in New York had a population of more than 150,000 that produced between three and four million pounds of manure and four thousand gallons of urine every day.[6] In contrast, a motor vehicle did not have to be fed, watered, or stabled. Upkeep was simpler, and travel was faster. Soon, day-to-day activities like sending messages, traveling, and transporting goods became challenges that were more easily solved without the horse.

An industrializing world found less and less value in the millions of animals that now occupied rural grasslands, including both domestic and feral horses. In North America, vast wild horse herds that had once acted as a valuable resource now became little more than grazing competition for American cattle ranches. Massive roundups began across the Great Plains and the Southern Cone, aimed at shipping feral mustangs by the cartload to factories and slaughterhouses that made use of horse meat or tendons for dog food and glue.[7] Horse populations plummeted in the United States from the tens of millions at the end of WWI down to the single-digit millions.

In the steppes of Eurasia, horse numbers remained more stable. In Mongolia, horse-based mail systems persisted well into the mid-20th century, and important cavalry battles were fought on horseback as late as 1950.[8] Nonetheless, the 20th century brought the golden age of horse transport to a dramatic close. Horses were downsized from military contexts around the world.[9] At the site of Swakopmund, in the deserts of Namibia, the South African Defence Force shot thousands of horses and mules suffering from an outbreak of disease.[10] Today, the bones of these animals still haunt the windswept Namibian dunes, an eerie memorial to the decline of horses in 20th-century warfare.

17 HOOFPRINTS

In much of the Western world today, horses have retreated far to the margins of daily life. Dirt roads have been paved, inns replaced with gas stations (and now, electric-vehicle charging ports). Outside of agriculture, tourism, or sport, horses are mostly invisible in most urban centers, except when police appear on horseback to control large crowds of protestors or partiers. In many places, particularly in the Western world, horse riding has become the purview of either the elite equestrian hobby rider or the rural rancher. Modern breeding practices have created a number of highly restricted lineages, dramatically reducing the genetic diversity of the world's horse populations.[1]

Despite horses' quick retreat from the industrial world, traces of their impact are everywhere, if you know where to look. And in some places, you need not look far. A visitor staying in Mongolia's urban capital of Ulaanbaatar around the height of the summer Naadam festival may still see a mounted rider making their way down a paved boulevard crammed with gridlocked traffic (see plate 20).

Even in the heart of the urban United States, horses are still deeply imbedded in culture, artwork, symbolism, and identity. Arriving at Denver International Airport, an enormous blue mustang with demonic eyes, nicknamed Blucifer, provides a startling welcome to visitors leaving the terminal to cheer on the Denver Broncos, another symbolic horse found almost everywhere across the state. Horse racing and equestrian sports are popular among many, and international competitions from the Kentucky Derby to the Olympics draw millions of viewers each year. Contemporary Western fashions include what is essentially horse equipment: jeans, trousers, and boots once worn by the riders of Inner Asia of centuries ago.

In our rapidly changing 21st-century world, the threads that link us to the past are increasingly important for navigating the future. As I write this book, I am sitting in my home along the eastern front of the Rocky Mountains in a house decorated with the gaudy horse brasses, horse art, and horsey furnishings of my Montana rancher forebears, Park and Ethel.

The world's transport infrastructure, although gradually mechanized from centuries before, still uses technologies, road systems, and traditions built outward from the horse. The design of our cars, roads, and trains—even the distance between our cities and towns—are traits inherited from the horse era. Horse paths have now become paved highways that carry motorized cars powered by petroleum derivatives amid a rapidly escalating crisis of emissions-linked climate change. My old Dodge pickup, a motorized horse carriage, sits in my driveway, resting from my commute along the road from Longmont to Boulder, the same road on which mail and passengers were first carried by horse and donkey in the 19th century.[2] Finding better transport systems for our future will require us to understand the strengths and weaknesses of the industrial replacements for horses and to find new ways to emulate the sustainability of older systems of horse travel in our modern solutions.

The tumult of modern geopolitics also continues to rework channels first formed during the horse era. As China invests billions in the Belt and Road Initiative, ancient travel corridors that once carried horses and silks are being rejuvenated to expand China's economic and political linkages across Eurasia. Social and economic networks shaped to a large degree by horse travel now influence the course of other transformative processes, like the spread of pandemic disease. Can a better understanding of ancient human-horse phenomena, like the formation of the Silk Road, help us understand modern trade and disease at a global scale? This may sound ambitious, but with better datasets and more partnerships between researchers and problem solvers, such a result is not out of the question.

Meanwhile, amid the chaos of development, globalization, and mechanization, horses remain a fountain of resilience, tradition, and connection. In horse country from Montana to Mongolia, horses continue in their oldest and arguably most important roles: companion, herd animal, conveyance for herders tending their flock. In Australia, Aboriginal ranchers and ranch hands maintain close relationships with horses at stations in the Outback. In the Americas, Indigenous peoples maintain and rekindle horse traditions as a source of community and healing. In Patagonia, rural *puesteros* still perform traditional hunts of wild rhea on horseback.[3] In the Great Plains, events like the Horse Nations Indian Relay bring together riders from across the continent, while many Native communities use horses in ceremonies, festivals, and memorials of important or tragic events. Members of the Lakota nation carefully conserve and care for traditional Native horse lineages, and horse riding serves as a powerful tool for youth programs. From the *buzkashi* of Afghanistan to the *pato* of Argentina and dozens of variants of rodeo, polo, and racing, horses are at the very center of sport and ceremony across continents.

Although solving the immense challenges of the 21st century is certainly beyond the current abilities of archaeology, new research and new technologies do hold the promise of a future where better knowledge of the past helps inform our modern problem solving. As laid down in the pages of this book, the applications of archaeological science to the study of horses have already revealed some astonishing things. With the aid of techniques such as ancient genomics and radiocarbon dating and better coverage of the archaeological record in historically underrepresented regions like the Eurasian steppes and the Great Plains, we now have a much clearer model for when horses were first domesticated, when transformative technologies like the saddle and stirrup were first developed, and when and how domestic horses were first integrated into societies across the ancient world.

Within archaeology itself, Indigenous archaeologists are working tirelessly to strip racism, bias, and colonial thought from the interpretation of horses and horse cultures in the archaeological record and to use archaeological scholarship to recognize the cultural achievements of Native horse nations in Western scientific literature. Renewed investment in understanding, prioritizing, and protecting these important relationships may help heal some of the wounds created by European colonization and the rapid removal of horses from the industrial world. With better data, we can start to test new ideas about *why* and *how* these transformative changes took place and explore aspects of the human-horse relationship that were never documented in texts. Increasing collaboration between anthropologists, historians, geneticists, paleoclimatologists, paleontologists, archaeological scientists, and Indigenous scholars from horse cultures from across our modern world will give us a better chance at tackling problems of the future armed with our knowledge of the past.

At Morin Mort in the Khangai Mountains, life continues much the same as it has since horse hoof images were first etched into its silent blue-stone panels more than three millennia ago. Although the occasional hum of a motorbike, the buzz of a cell phone, or the black outline of solar panels against a white yurt serve as reminders of the passage of the years, the patient visitor will still find the timeless image of a solitary rider watching over their herds in the fading evening light. Here, it seems, the link between people and horses, between past and present, is fully alive, hoof beats etched and echoing in the stone.

ILLUSTRATIONS

Plates follow page 120

NOTES

CHAPTER 1. EVOLUTION

1. MacFadden, *Fossil Horses*, 90.
2. Sussman et al., "Rethinking Primate Origins."
3. Wood et al., "Postcranial Functional Morphology."
4. Strömberg, "Evolution of Grasses."
5. MacFadden, "Fossil Horses."
6. Hildebrand, "The Mechanics of Horse Legs"; McHorse et al., "Mechanics of Evolutionary Digit Reduction."
7. Ransom and Kaczensky, *Wild Equids*.
8. Krueger, "Social Ecology of Horses."
9. Wit et al., *The Tale of the Przewalski's Horse*.
10. Krueger, "Social Ecology of Horses."
11. Krueger, "Social Ecology of Horses."
12. Brubaker et al., "Cognition and Learning in Horses."
13. Orlando et al., "Recalibrating Equus Evolution."
14. MacFadden, "Fossil Horses."
15. Rosenbom et al., "Reassessing the Evolutionary History of Ass-Like Equids."
16. Barrón-Ortiz et al., "What Is Equus"; Heintzman et al., "A New Genus of Horse from Pleistocene North America."
17. Vershinina et al., "Ancient Horse Genomes."
18. Fages et al., "Tracking Five Millennia of Horse Management."
19. Orlando et al., "Recalibrating Equus Evolution."

CHAPTER 2. CONNECTION

1. Orlando et al., "Recalibrating Equus Evolution."

2. Bernor et al., "Evolution of Early Equus in Italy"; Olsen, "The Role of Humans in Horse Distribution."

3. Bernor et al., "Evolution of Early Equus in Italy."

4. Pope et al., *The Horse Butchery Site.*

5. Pope et al., *The Horse Butchery Site.*

6. Vasiliev, "Large Mammal Fauna."

7. Straus, "Upper Paleolithic Hunting Tactics."

8. Burke et al., "Histological Observations of Cementum Growth."

9. Olsen, "Solutré."

10. Hoffecker et al., "The Hunting of Horse."

11. Hoffecker et al., "The Hunting of Horse"; Levine, "The Origins of Horse Husbandry."

12. Niven, "From Carcass to Cave."

13. Guthrie, "Human-Horse Relations."

14. Pruvost et al., "Genotypes of Predomestic Horses"; Guthrie, "Human-Horse Relations."

15. Pruvost et al., "Genotypes of Predomestic Horses."

16. Sauvet, "The Hierarchy of Animals."

17. Bourgeon et al., "Earliest Human Presence."

18. Dixon, *Bones, Boats and Bison.*

19. Davis et al., "Late Upper Paleolithic Occupation"; Webb et al., "Last Horses and First Humans."

20. McHorse et al., "What Species of Horse Was Coeval."

21. Waters et al., "Late Pleistocene Horse and Camel Hunting"; Kooyman et al., "Late Pleistocene Horse Hunting."

22. Brink et al., "Pleistocene Horse."

23. Webb et al., "Last Horses and First Humans."

24. Steele et al., "AMS 14C Dating"; Villavicencio et al., "Combination of Humans, Climate, and Vegetation Change."

25. Guthrie, "Rapid Body Size Decline."

26. Haile, "Ancient DNA."

27. Collin, "The Relationship between the Indigenous Peoples of the Americas and the Horse."

28. Orlando et al., "Recalibrating Equus Evolution."

29. Vasil'ev, "Faunal Exploitation."

30. Boeskorov et al., "A Study of a Frozen Mummy of a Wild Horse"; Fages et al., "Tracking Five Millennia of Horse Management."

31. Faith, "Late Pleistocene and Holocene Mammal Extinctions."

32. Fages et al., "Tracking Five Millennia of Horse Management."

33. Bendrey, "From Wild Horses to Domestic Horses."

34. Arbuckle, "Animals and Inequality"; Martin et al., "The Equid Remains"; Guimaraes et al., "Ancient DNA."

35. Bendrey, "From Wild Horses to Domestic Horses."

CHAPTER 3. TRACING DOMESTICATION

1. Koungoulos et al., "Hunting Dogs Down Under?"

2. Bergström et al., "Grey Wolf Genomic History."

3. Lahtinen, "Excess Protein."

4. Perri et al., "New Evidence"; Perri et al., "Dog Domestication."

5. Willcox et al., "Large-Scale Cereal Processing."

6. Daly et al., "Ancient Goat Genomes."

7. Zeder et al., "The Initial Domestication of Goats."

8. Zeder, "Out of the Fertile Crescent"; Taylor et al., "Evidence for Early Dispersal."

9. Vigne, "Early Domestication and Farming."

10. Benecke, "Late Prehistoric Exploitation."

11. Deng, "The Fossils of the Przewalski's Horse"; Foronova, "Late Quaternary Equids."

12. Tarasov et al., "A Continuous Late Glacial and Holocene Record"; Zhang et al., "Holocene Climate Variations"; Perșoiu et al., "Holocene Winter Climate Variability."

13. Guthrie, "Rapid Body Size Decline."

14. Steppan, "The Neolithic Human Impact."

15. Kosintsev, "The Human-Horse Relationship"; Rassamakin, "The Eneolithic of the Black Sea Steppe."

16. Kosintsev, "The Human-Horse Relationship."

17. Rassamakin, "The Eneolithic of the Black Sea Steppe."

18. Jeong et al., "A Dynamic 6,000-Year Genetic History"; Haak et al., "Massive Migration."

19. Gimbutas, *The Kurgan Culture.*

20. Benecke, "Late Prehistoric Exploitation"; Greenfield, "The Social and Economic Context."

21. Anthony et al., "Early Horseback Riding."

22. Anthony, *The Horse, the Wheel and Language.*

23. Anthony et al., "Early Horseback Riding"; Kuzmina, "Mythological Treatment."

24. Levine, "The Origins of Horse Husbandry."

25. Taylor, "Horse Demography"; Levine, "The Origins of Horse Husbandry."

26. Clutton-Brock, "The Buhen Horse."

27. Taylor et al., "Horseback Riding."

28. Clutton-Brock, "The Buhen Horse"; Anthony et al., "The Origins of Horseback Riding"; Takács, "Evidence of Horse Use."

29. Cook, "Damage by the Bit"; Taylor et al., "Reconstructing Equine Bridles."

30. Taylor et al., "Reconstructing Equine Bridles"; Takács, "Evidence of Horse Use."

31. Levine et al., "Abnormal Thoracic Vertebrae"; Li et al., "Early Evidence for Mounted Horseback Riding"; Taylor, Hart, et al., "Interdisciplinary Analysis"; Weber, "Elite Equids"; Levine, "Chinese Chariot Horses"; Taylor et al., "Understanding Early Horse Transport."

32. Taylor et al., "Early Dispersal of Domestic Horses"; Taylor et al., "Equine Cranial Morphology."

33. Taylor et al., "Equine Cranial Morphology."

34. Taylor et al., "Origins of Equine Dentistry"; Rannamäe et al., "A Month in a Horse's Life."

35. Makarewicz et al., "Isotopic Evidence"; Taylor et al., "Interdisciplinary Analysis"; Bendrey et al., "Patterns of Iron Age Horse Supply."

36. Gaunitz et al., "Ancient Genomes."

37. Anthony et al., "The Origin of Horseback Riding."

38. Anthony et al., "Eneolithic Horse Exploitation."

39. Horwitz et al., "A Late Neolithic Equid Offering."

40. Rassamakin. "The Eneolithic of the Black Sea Steppe."

41. Levine, "The Origins of Horse Husbandry."

42. Esin, "An Eastern Tibetan Tool"; Reinhold et al., "Contextualising Innovation."

43. Olsen, "The Exploitation of Horses"; Olsen, "Early Horse Domestication."

44. Olsen, "Early Horse Domestication"; French et al., "Geomorphological and Micromorphological Investigations."

45. Brown et al., "Bit Wear."

46. Levine, "Exploring the Criteria for Early Horse Domestication."

47. Brown et al., "Bit Wear"; Olsen, "Early Horse Domestication."

48. Benecke et al., "Horse Exploitation."

49. Levine, "The Origins of Horse Husbandry"; Olsen, "Early Horse Domestication"; Fages et al., "Horse Males Became Over-Represented."

50. Ingold, *Hunters, Pastoralists and Ranchers.*

51. Olsen, "Early Horse Domestication."

52. Outram et al., "The Earliest Horse Harnessing."

53. Bendrey, "New Methods"; Bendrey, "An Analysis of Factors."

54. Outram et al., "The Earliest Horse Harnessing."

55. Dudd et al., "Organic Residue Analysis."

56. Gaunitz et al., "Ancient Genomes."

57. Wit et al., *The Tale of the Przewalski's Horse*; Olsen, "Early Horse Domestication," 95.

58. Barros Damgaard et al., "The First Horse Herders."

59. Taylor et al., "Rethinking the Evidence."

60. Wilkin et al., "Dairying."

61. Scott et al., "Emergence and Intensification."

CHAPTER 4. WHEELS

1. Gaastra et al., "Gaining Traction."

2. Burmeister et al., "Some Notes on Pictograms."

3. Bakker et al., "The Earliest Evidence of Wheeled Vehicles."

4. Maurer et al., "Cattle Drivers from the North?"

5. Reinhold et al., "Contextualising Innovation."

6. Sagona, *The Archaeology of the Caucasus.*

7. Kohl, *The Making of Bronze Age Eurasia.*

8. Reinhold et al., "Contextualising Innovation."

9. Littauer et al., *Selected Writings on Chariots*; Shishlina et al., "Catacomb Culture Wagons."

10. Haak et al., "Massive Migration."

11. Allentoft et al., "Population Genomics"; Hermes et al., "Mitochondrial DNA."

12. Jeong et al., "A Dynamic 6,000-Year Genetic History."

13. Kovalev, *Earliest European*; Kovalev et al., "Discovery of New Cultures."

14. Wilkin et al., "Dairy Pastoralism."

15. Trautmann et al., "First Bioanthropological Evidence."

16. Taylor et al., "Investigating Reindeer Pastoralism."

17. Taylor et al., "Early Pastoral Economies."

18. Todd et al., "The Genomic History."

19. Kimura et al., "Ancient DNA."

20. Rossel et al., "Domestication of the Donkey"; Mitchell, *The Donkey in Human History*, 53.

21. Arnold et al., "Isotopic Evidence"; Greenfield et al., "Evidence for Movement of Goods."

22. Zarins et al., *The Domestication of Equidae*; Mitchell, *The Donkey in Human History*.

23. Bennett et al., "Taming the Late Quaternary Phylogeography."

24. Orlando et al., "Geographic Distribution."

25. Zarins et al., *The Domestication of Equidae*.

26. Zarins et al., *The Domestication of Equidae*. Here, the all-capital-letter transliteration of the original Sumerian logograms follows the terms used in Recht, *The Spirited Horse*.

27. Zarins et al., *The Domestication of Equidae*.

28. Zarins et al., *The Domestication of Equidae*.

29. Recht, *The Spirited Horse*.

30. Kuz'mina, *The Origin of the Indo-Iranians*, 134, 332.

31. Moorey, "Pictorial Evidence."

32. Van Buren, "Clay Figurines."

33. Kawami, "That Strange Equid."

34. Zarins et al., *The Domestication of Equidae*, 147–48.

35. Blench, "Wild Asses and Donkeys."

36. Littauer et al., *Wheeled Vehicles*; Drews, *Early Riders*.

37. Gilbert, "Equid Remains"; Bökönyi et al., "A Review of Animal Remains"; Vila, "Data on Equids."

38. Zeder, "The Equid Remains."

39. Shai et al., "The Importance of the Donkey"; Clutton-Brock et al., "More Donkeys from Tell Brak."

40. Gilbert, "Equid Remains."

41. Weber, "Elite Equids"; Weber et al., "Restoring Order."

42. Bennett et al., "The Genetic Identity."

43. Bennett et al., "The Genetic Identity."

44. Mühl, "'Metal Makes the Wheel Go Round'"; Littauer et al., *Selected Writings on Chariots*, 261–71.

45. Littauer et al., *Wheeled Vehicles*.

46. Littauer et al., *Selected Writings on Chariots*, 479–86.

47. Littauer et al., *Wheeled Vehicles*.

48. Littauer et al., *Wheeled Vehicles*.

49. Greenfield et al., "Earliest Evidence"; Clutton-Brock, "Were the Donkeys at Tell Brak (Syria) Harnessed"; Littauer et al., *Selected Writings on Chariots*; Bar-Oz et al., "Symbolic Metal Bit."

50. Littauer et al., *Wheeled Vehicles*.

51. Noble, "The Mesopotamian Onager."

52. Olsen et al., *A Gift from the Desert*, 95.

53. Drews, *Early Riders*.

CHAPTER 5. CHARIOTS

1. Librado et al., "The Origins and Spread of Domestic Horses."

2. Librado et al., "The Origins and Spread of Domestic Horses."

3. Librado et al., "The Origins and Spread of Domestic Horses"; Olsen, Early Horse Domestication."

4. Wilkin et al., "Dairying."

5. Wilkin et al., "Dairying."

6. Scott et al., "Emergence and Intensification of Dairying."

7. Der Sarkissian et al., "Evolutionary Genomics"; Librado et al., "The Origins and Spread of Domestic Horses."

8. Dietz, "Horseback Riding."

9. Dietz, "Horseback Riding."

10. Mitchell, "The Constraining Role of Disease."

11. Swart, *Riding High*.

12. Taylor, "How Dan the Zebra Stopped An Ill-Fated Government Breeding Program."

13. Swart, *Riding High*, 26.

14. Root, *The Overland Stage to California*, 50.

15. Root, *The Overland Stage to California*, 427.

16. Librado et al., "The Origins and Spread of Domestic Horses."

17. Chechushkov et al., "Relative and Absolute Chronologies"; Izbitser, "Kolyesnitsa."

18. Chechushkov et al., "Relative and Absolute Chronologies."

19. Lindner, "Chariots in the Eurasian Steppe."

20. Kuznetsov, "The Emergence of Bronze Age Chariots."

21. Chechushkov et al., "Eurasian Steppe Chariots."

22. Moorey, "The Emergence of the Light, Horse-Drawn Chariot."

23. Spengler et al., "An Imagined Past?"; Schmaus, "Animals, Households, and Communities."

24. Kosintsev, "The Human-Horse Relationship."

25. Kosintsev, "The Human-Horse Relationship."

26. Anthony et al., "Early Horseback Riding and Warfare," 137.

27. Wilkin et al., "Dairying"; Wilkin et al., "Dairy Pastoralism"; Scott et al., "Emergence and Intensification of Dairying."

28. Raulwing, *Horses, Chariots and Indo-Europeans*.

29. Crouwel et al., *Chariots and Related Equipment*.

30. Esin et al., "Chariots in the Bronze Age."

31. Jacobson-Tepfer, *The Hunter, the Stag, and the Mother of Animals*.

32. Jacobson-Tepfer, *The Hunter, the Stag, and the Mother of Animals*, 192, 204.

33. Esin et al., "Chariots in the Bronze Age."

34. Kuz'mina, *The Origin of the Indo-Iranians*, 34; Anthony, "The Sintashta Genesis."

35. Chechushkov et al., "Early Evidence for Horse Utilization."

36. Taylor et al., "Equine Cranial Morphology."

37. Librado et al., "The Origins and Spread of Domestic Horses."

38. Chechushkov et al., "Eurasian Steppe Chariots."

39. Svyatko et al., "New Radiocarbon Dates"; Poliakov et al., "Modern Data on the Bronze Age Radiocarbon Chronology."

40. Motuzaite Matuzeviciut et al., "Ecology and Subsistence."

41. Anthony, "The Sintashta Genesis."

42. Kuznetsov, "The Emergence of Bronze Age Chariots."

43. Kyselý et al., "Horse Size and Domestication"; Czebreszuk et al., "The Horse, Wagon, and Roads."

44. Goldhahn, "Bredarör on Kivik."

45. Bendrey, "The Horse."

46. Drews, *The Coming of the Greeks*; Raulwing, *Horses, Chariots and Indo-Europeans*; Marzahn, "Equids in Mesopotamia."

47. Zarins et al., *The Domestication of Equidae.*

48. Drews, *The Coming of the Greeks.*

49. Drews, *The Coming of the Greeks*, 98.

50. Kuz'mina, *The Origin of the Indo-Iranians*, 128.

51. Guimaraes et al., "Ancient DNA."

52. Littauer et al., *Wheeled Vehicles.*

53. Moorey, "The Emergence of the Light, Horse-Drawn Chariot" (specific descriptions of projectile-platform function and rapid troop deployment is on p. 204); Littauer et al., *Wheeled Vehicles.*

54. Josephus, *Against Apion.*

55. Drews, *The Coming of the Greeks*; Clutton-Brock, "The Buhen Horse"; Olsen et al., *A Gift from the Desert.*

56. Näser et al., "Of Kings and Horses."

57. Law, *The Horse in West African History*; Olsen et al., *A Gift from the Desert.*

58. Ikram et al., *Catalogue General*, 46–47.

59. Crouwel et al., *Chariots and Related Equipment.*

60. Badisches Landesmuseum Karlsruhe et al., "The Reconstruction of Pi-Ramesse."

61. Littauer et al.. *Wheeled Vehicles.*

62. Mitchell, *Horse Nations*; Mitchell, *The Donkey in Human History.*

63. Hamilakis, "A Footnote on the Archaeology of Power."

64. Harding, "Horse-Harness."

65. Maran et al., "A Horse-Bridle Piece"; Podobed et al., "Cheek-Pieces of the Water Horses."

66. Drews, *The Coming of the Greeks.*

67. Harding, "Horse-Harness"; Maran et al., "A Horse-Bridle Piece."

68. Lazaridis et al., "Genetic Origins of the Minoans and Mycenaeans."

69. Danino, "The Horse and the Aryan Debate"; Anthony, "Horse, Wagon & Chariot."

70. Danino, "The Horse and the Aryan Debate."

71. Kumar, "A Note on Chariot Burials."

72. Kuz′mina, *The Origin of the Indo-Iranians*, 336–37.

73. Kenoyer, "Cultures and Societies of the Indus Tradition."

74. Zahir, "Gandhara Grave Culture"; Muhammad Zahir, email communication with the author, July 2020.

75. Jawad, "Faunal Remains."

76. Zahir, "Gandhara Grave Culture"; Agrawal et al., "Cist Burials."

77. Azzaroli, "Two Proto-Historic Horse Skeletons."

78. Azzaroli, "Two Proto-Historic Horse Skeletons."

79. Narasimhan et al., "The Formation of Human Populations."

80. Shetenko, "Время Появления Домашней Лошади" [The Time of the Appearance of the Domesticated Horse].

81. Narasimhan et al., "The Formation of Human Populations."

82. Kuz′mina, *The Origin of the Indo-Iranians*.

83. Drews, *The Coming of the Greeks*.

84. Kuz′mina, *The Origin of the Indo-Iranians*; Anthony, "The Roles of Climate Change," 47.

85. Kuz′mina, *The Origin of the Indo-Iranians*; Spengler et al., "Early Agriculture and Crop Transmission"; Miller et al., "Millet Cultivation across Eurasia."

CHAPTER 6. ORACLE BONES

1. Li et al., "Heading North."

2. Massilani et al., "Denisovan Ancestry"; Li et al., "Heading North."

3. Cai et al., "Radiocarbon and Genomic Evidence."

4. Jeong et al., "A Dynamic 6,000-Year Genetic History."

5. Jeong et al., "A Dynamic 6,000-Year Genetic History."

6. Losey et al., "A Second Mortuary Hiatus"; Taylor et al., "Early Pastoral Economies."

7. Houle et al., "Resilient Herders."

8. Taylor et al., "Early Pastoral Economies."

9. Taylor et al., "Early Pastoral Economies."

10. Svyatko et al., "New Radiocarbon Dates"; Poliakov et al., "Modern Data on the Bronze Age Radiocarbon Chronology."

11. Svyatko et al., "New Radiocarbon Dates."

12. Legrand, "The Emergence of the Scythians."

13. Legrand, "The Emergence of the Scythians."

14. Fitzhugh, "The Mongolian Deer Stone-Khirigsuur Complex."

15. Esin et al., "Paint on Deer Stones of Mongolia."

16. Bayarsaikhan, Монголын умард нутгийн буган хөшөөдд [Deer Stones of Northern Mongolia]; Tabaldyev, "Monuments of the Bronze Age of Kyrgyzstan."

17. Seitsonen et al., "The Mystery of the Missing Caprines."

18. Taylor et al., "Horse Sacrifice and Butchery."

19. Allard et al., "Khirigsuurs, Ritual and Mobility."

20. Taylor et al., "Horse Sacrifice and Butchery."

21. Taylor et al., "Horse Sacrifice and Butchery."

22. Lepetz et al., "Customs, Rites, and Sacrifices"; Lazzerini et al., "Season of Death of Domestic Horses."

23. Zazzo et al., "High-Precision Dating."

24. Fages et al., "Tracking Five Millennia of Horse Management."

25. Taylor et al., "Equine Cranial Morphology"; Taylor et al., "Reconstructing Equine Bridles"; Taylor et al., "Horseback Riding."

26. Taylor, "Horse Demography."

27. Fitzhugh, "The Mongolian Deer Stone-Khirigsuur Complex," 77

28. Taylor, "Horse Demography."

29. Allard et al., "Ritual Horses"; Taylor et al., "Understanding Early Horse Transport"; Lepetz et al., "Customs, Rites, and Sacrifices."

30. Taylor et al., "Understanding Early Horse Transport."

31. Taylor et al., "Horseback Riding."

32. Taylor et al., "Understanding Early Horse Transport."

33. Taylor et al., "Understanding Early Horse Transport."

34. Wagner et al., "Radiocarbon-Dated Archaeological Record."

35. Turfan City Bureau of Cultural Relics et al., "新疆洋海墓地" [Report of Archaeological Excavations at Yanghai Cemetery].

36. Beck et al., "The Invention of Trousers."

37. Wertmann et al., "New Evidence for Ball Games in Eurasia."

38. Cultural Relics Team of Xinjiang Museum, "且末县扎滚鲁克五座墓葬发掘简报" [Excavation Report on Five Burials].

39. Taylor et al., "Early Pastoral Economies."

40. Houle, "Emergent Complexity"; Wilkin et al., "Dairy Pastoralism."

41. Taylor et al., "Origins of Equine Dentistry."

42. Allard et al., "Khirigsuurs, Ritual and Mobility."

43. Taylor et al., "Equine Cranial Morphology."

44. Karstens et al., "A Palaeopathological Analysis."

45. Eregzen, *Ancient Funeral Monuments*; Taylor et al., "A Bayesian Chronology."

46. Cai et al., "Radiocarbon and Genomic Evidence."

47. Yuan et al., "Research on Early Horse Domestication"; Mair, "The Horse in Late Prehistoric China."

48. Janz et al., "Expanding Frontier and Building the Sphere."

49. Rawson et al., "Chariotry and Prone Burials."

50. Wu, *Chariots in Early China*.

51. Cooke, *Imperial China*, 30.

52. Rawson et al., "Seeking Horses."

53. Taylor et al., "Equine Cranial Morphology"; Wu, *Chariots in Early China*.

54. Piggott, "Chariots in the Caucasus and in China"; Littauer et al., *Selected Writings on Chariots*, 106-35.

55. Cooke, *Imperial China*.

56. Rawson et al., "Seeking Horses."

57. Mair, "The Horse in Late Prehistoric China."

58. Kuz'mina, *The Origin of the Indo-Iranians*.

59. Mair, "The Horse in Late Prehistoric China"; Bjørn, "Indo-European Loanwords."

60. Mair, "The Horse in Late Prehistoric China."

61. Rawson et al., "Seeking Horses."

62. Wu, *Chariots in Early China*, 7-9.

63. Rawson et al., "Seeking Horses."

CHAPTER 7. HORSEBACK

1. Outram et al., "Horses for the Dead."

2. Littauer et al., *Wheeled Vehicles*, 66-68.

3. Littauer et al., *Wheeled Vehicles*.

4. Schulman, "Egyptian Representations of Horsemen."

5. Littauer, "Bits and Pieces."

6. Taylor et al., "Understanding Early Horse Transport."

7. Bar-Oz et al., "Symbolic Metal Bit and Saddlebag Fastenings."

8. Littauer et al., *Selected Writings on Chariots,* 505–18.

9. Librado et al., "The Origins and Spread of Domestic Horses."

10. Sagona, *The Archaeology of the Caucasus*; Olsen et al., *A Gift from the Desert*, 103; Littauer et al., *Selected Writings on Chariots,* 333.

11. Stepanov et al., "Horse Equipment."

12. Taylor et al.,"Reconstructing Equine Bridles"; Taylor et al., "Understanding Early Horse Transport."

13. Khorvat, "Захоронения коней в камере № 31 Кургана Аржан-1" [Horse Burials in Chamber No. 31 of Kurgan Arzhan-1].

14. Benecke, "The Horse Skeletons."

15. Anglim et al., *Fighting Techniques.*

16. Anglim et al., *Fighting Techniques.*

17. Kuz'mina, *The Origin of the Indo-Iranians,* 367–76.

18. Schauensee, "Horse Gear from Hasanlu."

19. Gnecchi-Ruscone et al., "Ancient Genomic Time Transect"; Jeong et al., "A Dynamic 6,000-Year Genetic History."

20. Caspari et al., "Tunnug 1 (Arzhan 0)"; Sadykov et al., "Kurgan Tunnug 1"

21. Librado et al., "Ancient Genomic Changes."

22. Librado et al., "Ancient Genomic Changes."

23. Benecke, "The Horse Skeletons."

24. Lepetz, "Horse Sacrifice"; Vitt, "The Horses of the Kurgans of Pazyryk"; Li et al., "Early Evidence for Mounted Horseback Riding."

25. Liu et al., "A Single-Nucleotide Mutation."

26. Kosintsev, "The Human-Horse Relationship."

27. Taylor et al., "Early Pastoral Economies."

28. Wilkin et al., "Dairy Pastoralism."

29. Tuvshinjargal et al., "Гэрийн Эртний Түүхэн" [Clarifying the Early Historical Development of the Ger]; Andrews, *Felt Tents and Pavilions.*

30. Baumer, *The History of Central Asia,* 225.

31. Mlinar et al., *Hear the Horses of Celts.*

32. Seguin-Orlando et al., "Heterogeneous Hunter-Gatherer."

33. Baumer, *The History of Central Asia.*

34. Mokrynin, Археология и история древнего и средневекового Кыргызстана: избранное [Archaeology and History of Ancient and Medieval Kyrgyzstan].

35. Azzaroli, "Two Proto-Historic Horse Skeletons."

36. Honeychurch, "Inner Asia and the Spatial Politics of Empire," 128.

37. Mayor, *The Amazons.*

CHAPTER 8. HORSE PEOPLE

1. Mayor, *The Amazons.*

2. Spengler et al., "An Imagined Past?"

3. Spengler et al., "An Imagined Past?"

4. Stark et al., *Nomads and Networks,* 121.

5. Lepetz et al., "To Accompany and Honour the Deceased."

6. Rudenko, *Frozen Tombs of Siberia.*

7. Lepetz et al., "To Accompany and Honour the Deceased."

8. Rudenko, *Frozen Tombs of Siberia.*

9. Taylor et al., "Origins of Equine Dentistry."

10. Wertmann et al., "The Earliest Directly Dated Saddle."

11. Stepanova et al., "Horse Equipment."

12. Wertmann et al., "No Borders for Innovations."

13. Bennett, *Conquerors.*

14. Minetti, "Physiology."

15. Olsen et al., *A Gift from the Desert,* 36.

16. Xenophon, *The Art of Horsemanship.*

17. Crouwel, *Chariots and Other Wheeled Vehicles.*

18. Corbino et al., "Equine Exploitation at Pompeii."

19. Wagner et al., "The Ornamental Trousers from Sampula."

20. Rawson et al., "Seeking Horses."

21. Cooke, *Imperial China,* 40-41.

22. Li et al., "Horses in Qin Mortuary Practice."

23. Cooke, *Imperial China.*

24. Beckwith, *Empires of the Silk Road.*

25. Kradin, "Stateless Empire."

26. Iderkhangai et al., "Хүннүгийн Лунчэн, Чанюйтин, Лунтин, хэмээх үгсийн тухай, Луут хот хэмээн бичигдэх болсон шалтгаан" [About the Xiongnu Period Words Longcheng, Chanyuting, and Longting].

27. Miller et al., "Proto-Urban Establishments."

28. Erdene-Ochir et al., Ноён уулын дурсгалын археологийн шинэ судалгаа [New Archaeological Research at the Site of Noyon Uul], 206.

29. Erdenebaatar, "Material Cultural Heritage of Xiongnu Empire."

30. Wilkin et al., "Economic Diversification"; Wright et al., "The Xiongnu Settlements."

31. Jeong et al., "A Dynamic 6,000-Year Genetic History."

32. Chan, *Nomadic Empires*, 82.

33. Bayarsaikhan et al., "The Origins of Saddle and Riding Technology."

34. Turbat et al., "Xiongnu Archaeological Tamgas."

CHAPTER 9. THE SILK AND TEA ROADS

1. Creel, "The Role of the Horse in Chinese History."

2. Shelach-Lavi et al., "Cavalry and the Great Walls."

3. Cooke, *Imperial China*; Liu, "Migration and Settlement."

4. Cooke, *Imperial China*, 41.

5. Jones, "Wings across the Silk Road."

6. Kuz'mina, *The Origin of the Indo-Iranians*.

7. Cooke, *Imperial China*.

8. Hyland, *Equus*.

9. Guedes et al., "The Archaeology of the Early Tibetan Plateau."

10. Aldenderfer, "Variation in Mortuary Practice."

11. Zhang et al., "Identification and Interpretation of Faunal Remains."

12. Yang et al., "Haplotype Diversity."

13. Tao et al., "The Coffin Paintings of the Tubo Period."

14. Lu et al., "Earliest Tea."

15. Hoh et al., *The True History of Tea*.

CHAPTER 10. STEPPE EMPIRES

1. Dean, "A Descriptive Label for Spurs"; De Lacy, *History of the Spur*.

2. De Lacy, *The History of the Spur*.

3. Littauer et al., *Selected Writings on Chariots*; Law, *The Horse in West African History*.

4. Bayarsaikhan et al., "The Origins of Saddle and Riding Technology."

5. Stepanova, "Saddles of the Hun-Sarmatian Period."

6. Dien, "The Stirrup"; Bayarsaikhan et al., "The Origins of Saddle and Riding Technology."

7. Dien, "The Stirrup."

8. Bayarsaikhan et al., "The Origins of Saddle and Riding Technology."

9. Dien, "The Stirrup."

10. Stepanova, "Saddles of the Hun-Sarmatian Period."

11. Bayarsaikhan et al., "The Origins of Saddle and Riding Technology."

12. Caprioli, "Equestrian Military Equipment"; Curta, "The Earliest Avar-Age Stirrups."

13. Struck et al., "Climate Change and Equestrian Empires."

14. Pederson et al., "Pluvials, Droughts"; Putnam et al., "Little Ice Age Wetting."

15. Su et al., "Impact of Climate Change"; McCormick et al., "Climate Change."

16. Harbeck et al., "Yersinia Pestis DNA."

17. Kausrud et al., "Modeling the Epidemiological History of Plague."

18. Rogers et al., "Urban Centres."

19. Fenner et al., "Stable Isotope and Radiocarbon Analyses."

20. Shim, "The Postal Roads of the Great Khans."

21. Minetti, "Physiology"; Atwood, "Mongol Messenger's Badge."

22. Atwood, "Mongol Messenger's Badge."

23. Weatherford, *Genghis Khan*.

24. Librado et al., "Tracking the Origins of Yakutian Horses"; Cooper et al., "Evidence of Eurasian Metal Alloys."

CHAPTER 11. DESERT AND SAVANNA EMPIRES

1. Morgan et al., "The Effect of Coat Clipping."

2. Olsen et al., *A Gift from the Desert*, 54.

3. Macdonald, "Hunting, Fighting, and Raiding."

4. Macdonald, "Hunting, Fighting, and Raiding"; Olsen et al., *A Gift from the Desert*.

5. Mallory-Greenough, "The Horse Burials of Nubia"; Doxey, "Napatan Horses"; Schrader et al., "Symbolic Equids."

6. Kelekna, "Northern Africa."

7. Olsen et al., *A Gift from the Desert*, 18.

8. Olsen et al., *A Gift from the Desert*, 18.

9. Robin, "Sabean and Himyarites Discover the Horse."

10. Olsen et al., *A Gift from the Desert*; Macdonald, "Hunting, Fighting, and Raiding."

11. Kelekna, *The Horse in Human History*.

12. Fages et al., "Tracking Five Millennia of Horse Management."

13. MacEachern et al., "Early Horse Remains."

14. Law, *The Horse in West African History*; Kefena et al., "Morphological Diversities."

15. Law, *The Horse in West African History*.

16. Law, *The Horse in West African History*, 395.

17. Rivallain, "The Horse."

18. Rivallain, "The Horse."

19. Law, *The Horse in West African History*, 123.

20. Law, *The Horse in West African History*.

21. Ogundiran, *The Yoruba*, 269.

22. Ogundiran, "The Formation of an Oyo Imperial Colony."

23. Diakakis et al., "Correlation between Equine Colic and Weather Changes."

24. Mitchell, "The Constraining Role of Disease."

25. Dennis et al., "Diseases May Shape the Distribution of Equid Species."

CHAPTER 12. OUT TO SEA

1. Kaniewsk et al., "The Sea Peoples."

2. D'Amato et al., *Sea Peoples*, 40–46.

3. Littauer et al., *Selected Writings on Chariots*, 141–73.

4. Hyland, *Equus*, 98–100.

5. Barnes, "The Emergence and Expansion of Silla."

6. Habu, "Seafaring."

7. Chiga, "日本に伝えられた馬文化" [Horse Culture Passed Down to Japan].

8. Chiga, "日本に伝えられた馬文化" [Horse Culture Passed Down to Japan].

9. Sasaki, "Adoption of the Practice of Horse-Riding."

10. Habu, "Seafaring" 167.

11. Chiga, "日本に伝えられた馬文化" [Horse Culture Passed Down to Japan].

12. Uetsuki et al., "Horse Feeding Strategy."

13. Tozaki et al., "Genetic Diversity"; Uetsuki et al., "The Use of Horses."

14. Skvorstov, "Burials of Riders and Horses"; Karczewska, "The Role of Horse Burials."

15. Dobat et al., "The Four Horses."

16. Dobat et al., "The Four Horses."

17. Kaliff et al., *The Great Indo-European Horse Sacrifice*; Hayhurst, "A Recent Find"; Ó Súilleabháin, "Foundation Sacrifices"; Hukantaival, "Horse Skulls."

18. Lepetz et al., "Historical Management of Equine Resources."

19. Reich, "The Cemetery of Oberhof"; Bliujienė et al., "Burials with Horses."

20. Carver, *Sutton Hoo*.

21. Klæsøe, *Viking Trade and Settlement*.

22. Nordeide, "The Oseberg Ship Burial."

23. Price, "The Vikings in Spain."

24. Fages et al. "Tracking Five Millennia of Horse Management."

25. Pilø et al., "Crossing the Ice."

26. Biknevicius et al., "Locomotor Mechanics."

27. Wutke et al., "The Origin of Ambling Horses."

28. Williams, *Weapons of the Viking Warrior*.

29. Pedersen, "Riding Gear."

30. Ingstad et al., *The Viking Discovery of America*.

31. Shenk, *To Valhalla by Horseback?*

32. Löffelmann et al., "Sr Analyses."

33. Kalliff et al., *The Great Indo-European Horse Sacrifice*; Nistelberger et al., "Sexing Viking Age Horses."

34. Shenk, *To Valhalla by Horseback?*, 16.

35. McGovern et al., "Zooarchaeology of the Scandinavian Settlements"; Smiarowski, "Climate-Related Farm-to-Shieling Transition."

36. Enghoff, *Hunting, Fishing and Animal Husbandry*, 77; Levine, "The Origins of Horse Husbandry."

37. Kuitems et al., "Evidence for European Presence"; Wallace, "L'Anse Aux Meadows."

38. Taylor et al., "Early Dispersal of Domestic Horses."

39. Meyer et al., "Pferdetransport zur See"; Pryor, "Transportation of Horses by Sea."

CHAPTER 13. THE RETURN

1. Crosby, *Ecological Imperialism*.

2. Bento et al., *History of the Azores*.

3. Bento et al., *History of the Azores*.

4. Deagan et al., *Columbus's Outpost*.

5. Street, "Feral Animals in Hispaniola."

6. Cabrera, *Caballos de América*, 113.

7. Delsol et al., "Analysis of the Earliest Complete mtDNA Genome."

8. Strassnig, "Rediscovering the Camino Real of Panama."

9. Cabrera, *Caballos de América*.

10. Olsen, "The Role of Humans in Horse Distribution," 115; Cabrera, *Caballos de América*.

11. Mitchell, *Horse Nations*, 77.

12. Cabrera, *Caballos de América*; Hudson, *Knights of Spain*, 74, 378.

13. Haile, *Jamestown Narratives*.

14. Chard, "Did the First Spanish Horses Landed in Florida and Carolina Leave Progeny?"

15. Cabrera, *Caballos de América*, 102.

16. Griggs, *The Archaeology of Central Caribbean Panama*, 160–63; 201.

17. Renton, *A Social and Environmental History*.

18. Mitchell, *Horse Nations*, 77.

19. Forbes, "The Appearance of the Mounted Indian."

20. Forbes, "The Appearance of the Mounted Indian."

21. Olsen, "The Role of Humans in Horse Distribution."

22. Olsen, "The Role of Humans in Horse Distribution," 115.

23. Forbes, "The Appearance of the Mounted Indian ."

24. Gifford-Gonzalez et al., "Foodways on the Frontier."

25. Haines, "The Northward Spread of Horses"; Roe, *The Indian and the Horse*, 8–9; Taylor et al., "Early Dispersal of Domestic Horses."

26. Taylor et al., "Early Dispersal of Domestic Horses"; Sundstrom, "Coup Counts and Corn Caches," 126.

27. Haile, *Jamestown Narratives*.

28. Jones, "The Old French-Canadian Horse."

29. Taylor et al., "Early Dispersal of Domestic Horses."

30. Holder, *The Hoe and the Horse on the Plains*, 80; Hämäläinen, *The Comanche Empire*.

31. Wilson, *The Horse and the Dog in Hidatsa Culture*.

32. Wayland et al., *Playing Cards of the Apaches*.

33. Johnson, *Lubbock Lake*, 149–50.

34. Wilson, *The Horse and the Dog in Hidatsa Culture*.

35. Mitchell, "'A Horse Race Is the Same All the World Over.'"

36. Carlson, *Eighteenth Century Navajo Fortresses*; Wedel, "Coronado, Quivira, and Kansas."

37. Wilson, *The Horse and the Dog in Hidatsa Culture*.

38. Catlin, *North American Indians*.

39. Mitchell, "Tracing Comanche History."

40. Ewers, *The Horse in Blackfoot Indian Culture*.

41. Reed, "Horses in Pawnee History and Culture."

42. Ewers, *The Horse in Blackfoot Indian Culture*.

43. Cowdrey et al., *American Indian Horse Masks*.

44. Ewers, *The Horse in Blackfoot Indian Culture*, 286–87

45. Grinnell, *Pawnee Hero Stories*; O'Shea, *Archaeology and Ethnohistory of the Omaha Indians*.

46. Wallace et al., *The Comanches*, 152.

47. Wheat et al., "The Olsen-Chubbuck Site."

48. Roe, *The Indian and the Horse*, 231.

49. Ewers, *The Horse in Blackfoot Indian Culture*, 154–68.

50. Mitchell, *Horse Nations*, 145.

51. Philipps, *Wild Horse Country*, 38.

52. Wallace et al., *The Comanches*, 119.

53. Taylor et al., "Early Dispersal of Domestic Horses."

54. Sundstrom, "Coup Counts and Corn Caches."

55. Schroeder, "The Alcova Redoubt"; Drass et al., "Digging Ditches."

56. Hämäläinen, *The Comanche Empire*.

57. Hämäläinen, *The Comanche Empire*; Hämäläinen, *Lakota America*.

CHAPTER 14. PAMPAS

1. Mitchell, *Horse Nations*, 220.

2. Bennett, *Conquerors*, 237–38.

3. deFrance, "Diet and Animal Use."

4. Kennedy et al., "Zooarchaeology and Changing Food Practices."

5. Fan et al., "Genomic Analysis."

6. deFrance, "Iberian Foodways."

7. Sauer, *The Archaeology and Ethnohistory of Araucanian Resilience*.

8. Mitchell, *Horse Nations*, 222.

9. Sauer, *The Archaeology and Ethnohistory of Araucanian Resilience*, 93–115.

10. Mitchell, *Horse Nations*, 258; Cabrera, *Caballos de América*.

11. Sauer, *The Archaeology and Ethnohistory of Araucanian Resilience*.

12. Gallardo et al., "Riders on the Storm"; Troncoso et al., "Making Rock Art."

13. Jong et al., "Mortuary Rituals"; Mitchell, *Horse Nations*, 260.

14. Camphora, *Animals and Society in Brazil*, 105.

15. Vander Velden, "A Tapuya 'Equestrian Nation?'"

16. Mitchell, *Horse Nations*, 236.

17. Camphora, *Animals and Society in Brazil*, 108; Vander Velden, "A Tapuya 'Equestrian Nation?'" 97.

18. Martinić Beros, *Los Aónikenk*, 74.

19. Taylor et al., "Early Domestic Horse Exploitation."

20. Mitchell, *Horse Nations*, 284.

21. Borrero et al., "Fragmented Records"; Belardi et al., "Late Holocene Guanaco Hunting Grounds"; Moreno et al., "Rastreando Ausencias."

22. Giardina et al., "Hunting, Butchering and Consumption of Rheidae."

23. Mazzanti et al., "Estrategias de Subsistencia"; Navarro, "Análisis Arqueofaunistico."

24. Taylor et al., Early Domestic Horse Exploitation; Musters, *At Home with the Patagonians*.

25. Guinnard, *Three Years' Slavery among the Patagonians*, 182.

26. Musters, *At Home with the Patagonians*, 131.

27. Mitchell, *Horse Nations*, 272.

28. Musters, *At Home with the Patagonians*, 76, 180; 140–41.

29. Guinnard, *Three Years' Slavery among the Patagonians*, 154; Musters, *At Home with the Patagonians*, 177.

30. Darwin, *Journal of Researches*, 39; Mitchell, *Horse Nations*, 283.

31. Navarro, "Análisis Arqueofaunistico"; Musters, *At Home with the Patagonians*.

32. Martinić, *Los Aónikenk*, 217, 221; Mitchell, *Horse Nations*, 239–58.

33. Bourne, *The Giants of Patagonia*; Mitchell, *Horse Nations*, 279.

34. Musters, *At Home with the Patagonians*, 130–31, 169.

35. Guinnard, *Three Years' Slavery among the Patagonians*, 196; Musters, *At Home with the Patagonians*, 77.

36. Guinnard, *Three Years' Slavery among the Patagonians*.

37. Guinnard, *Three Years' Slavery among the Patagonians*, 152.

38. Musters, *At Home with the Patagonians*, 177.

39. Mitchell, *Horse Nations*, 290.

40. Delaunay et al., "Glass and Stoneware Knapped Tools."

41. Guinnard, *Three Years' Slavery among the Patagonians*.

CHAPTER 15. INTO THE PACIFIC AND DOWN UNDER

1. Clarence-Smith, "Breeding and Power."

2. O'Connor et al., *Forts and Fortification*; Clarence-Smith, "Breeding and Power."

3. Clarence-Smith, "Breeding and Power," 40–41.

4. Swart, *Riding High*, 32.

5. Swart, *Riding High*, 40; Mitchell, *Horse Nations*.

6. Mallen et al., "The Rock Arts of Metolong."

7. Swart, *Riding High*, 80–102.

8. Blench, "The Austronesians in Madagascar"; O'Connor et al., *Forts and Fortification*.

9. Amano et al., "Archaeological and Historical Insights."

10. Clarence-Smith, "Elephants, Horses, and the Coming of Islam."

11. Fillios et al., "Who Let the Dogs In?"; Letnic et al., "Could Direct Killing by Larger Dingoes Have Caused the Extinction."

12. Crosby, *Ecological Imperialism*.

13. Schaffer, *Land Musters*.

14. Glover, *Archaeology in Eastern Timor*; Glover, "Excavations in Timor."
15. O'Connor et al., *Forts and Fortification*, 33.
16. Clements, *The Black War*.
17. Clements, *The Black War*, 81, 87.
18. Fijn, "Encountering the Horse," 12; Richards, *The Secret War*.
19. Sandall, *Coniston Muster*.
20. Fijn, "Encountering the Horse"; Jones, *Ochre and Rust*, 103.
21. Forbes, *Australia on Horseback*, 145.
22. Fijn, "Encountering the Horse"; Forbes, *Australia on Horseback*, 145.
23. Petrie, "Satisfaction in a Horse."
24. Mitchell, *Horse Nations*, 337.
25. Bergin et al., *The Hawaiian Horse*, 38–39.
26. Bergin et al., *The Hawaiian Horse*, 52.
27. Bergin et al., *The Hawaiian Horse*.
28. Bergin et al., *The Hawaiian Horse*.
29. Shackleton, *The Heart of the Antarctic*.

CHAPTER 16. IRON HORSES

1. Crisman et al., *When Horses Walked on Water*.
2. Root et al., *The Overland Stage to California*.
3. Root et al., *The Overland Stage to California*.
4. Taylor et al., "Early Dispersal of Domestic Horses."
5. Singleton, "Britain's Military Use of Horses."
6. Morris, "From Horse Power to Horsepower."
7. Philipps, *Wild Horse Country*; Bennett, *Conquerors*.
8. Taylor, "Pandemics and the Post"; Bessac et al., *Death on the Chang Tang*.
9. Law, *The Horse in West African History*, 204–6.
10. Swart, *Riding High*.

CHAPTER 17. HOOFPRINTS

1. Orlando, "Ancient Genomes."
2. Root et al., *The Overland Stage to California*, 554.
3. Giardina et al., "Hunting, Butchering and Consumption of Rheidae."

BIBLIOGRAPHY

Agrawal, D. P., Kharakwal, J., Kusumgar, S. and Yadava, M. G. Cist Burials of the Kumaun Himalayas. *Antiquity*, vol. 69, no. 264, pp. 550–54, September 1995.

Aldenderfer, M. Variation in Mortuary Practice on the Early Tibetan Plateau and the High Himalayas. *Journal of the International Association for Bon Research*, vol. 1, pp. 293–318, 2013.

Allard, F. and Erdenebaatar, D. Khirigsuurs, Ritual and Mobility in the Bronze Age of Mongolia. *Antiquity*, vol. 79, no. 305, pp. 547–63, September 2005.

Allard, F., Erdenebaatar, D., Olsen, S., Cavalla, A. and Maggiore, E. Ritual Horses in Bronze Age and Present Day Mongolia: Some Preliminary Observations from Khanuy Valley. In *Social Orders and Social Landscapes*, L. Popova, C. Hartley, and A. Smith, Eds., pp. 151–62. Cambridge Scholars Publishing, 2008.

Allentoft, M. E., Sikora, M., Sjögren, K.-G., Rasmussen, S., Rasmussen, M., Stenderup, J., Damgaard, P. B., et al. Population Genomics of Bronze Age Eurasia. *Nature*, vol. 522, no. 7555, pp. 167–72, June 11, 2015.

Amano, N., Bankoff, G., Findley, D. M., Barretto-Tesoro, G. and Roberts, P. Archaeological and Historical Insights into the Ecological Impacts of Pre-Colonial and Colonial Introductions into the Philippine Archipelago. *Holocene*, vol. 31, no. 2, pp. 313–30, February 1, 2021.

Andrews, P. A. *Felt Tents and Pavilions: The Nomadic Tradition and Its Interaction with Princely Tentage*. Melisende, 1999.

Anglim, S., Rice, R. S., Jestice, P., Rusch, S. and Serrati, J. *Fighting Techniques of the Ancient World (3000 B. C. to 500 A. D.): Equipment, Combat Skills, and Tactics*. Macmillan, 2003.

Anthony, D. W. *The Horse, the Wheel and Language: How Bronze-Age Riders from the Steppes Shaped the Modern World*. Princeton University Press, 2007.

Anthony, D. W. Horse, Wagon and Chariot: Indo-European Languages and Archaeology. *Antiquity*, vol. 69, no. 264, pp. 554–65, September 1995.

Anthony, D. W. The Roles of Climate Change, Warfare, and Long-Distance Trade. In *Social Complexity in Prehistoric Eurasia: Monuments, Metals and Mobility*, B. Hanks and K. Linduff, Eds., pp. 47–73. Cambridge University Press, 2009.

Anthony, D. W. and Brown, D. R. Eneolithic Horse Exploitation in the Eurasian Steppes: Diet, Ritual and Riding. *Antiquity*, vol. 73, no. 283, pp. 75–86, March 2000.

Anthony, D. W., Brown, D. R. and George, C. Early Horseback Riding and Warfare: The Importance of the Magpie around the Neck. In *Horses and Humans: The Evolution of Human-Equine Relationships*, S. Olsen, S. Grant, A. Choyke, and L. Bartosiewicz, Eds, pp. 137–56. British Archaeological Reports, 2006.

Anthony, D. W., Telegin, D. Y. and Brown, D. The Origin of Horseback Riding. *Scientific American*, vol. 265, no. 6, pp. 94–101, 1991.

Arbuckle, B. S. Animals and Inequality in Chalcolithic Central Anatolia. *Journal of Anthropological Archaeology*, vol. 31, no. 3, pp. 302–13, September 1, 2012.

Arnold, E. R., Hartman, G., Greenfield, H. J., Shai, I., Babcock, L. E. and Maeir, A. M. Isotopic Evidence for Early Trade in Animals between Old Kingdom Egypt and Canaan. *PloS One*, vol. 11, no. 6, p. e0157650, June 20, 2016.

Atwood, C. P. Mongol Messenger's Badge (Paiza or Gerege) in Pakpa Script. Project Himalayan Art, June 26, 2023. https://projecthimalayanart .rubinmuseum.org/essays/mongol-messengers-badge-paiza-or-gerege -in-pakpa-script/.

Azzaroli, A. Two Proto-Historic Horse Skeletons from Swāt, Pakistan. *East and West*, vol. 25, no. 3/4, pp. 353–57, 1975.

Badisches Landesmuseum Karlsruhe, Roemer-Pelizaeus Museum Hildesheim and Qantir Excavation Project. The Reconstruction of Pi-Ramesse. Artefacts—Scientific Illustration and Archaeological Reconstruction, 2016. https://www.artefacts-berlin.de/portfolio-item/the-reconstruction-of-pi-ramesse/.

Bakker, J. A., Kruk, J., Lanting, A. E. and Milisauskas, S. The Earliest Evidence of Wheeled Vehicles in Europe and the Near East. *Antiquity*, vol. 73, no. 282, pp. 778–90, December 1999.

Barnes, G. L. The Emergence and Expansion of Silla from an Archaeological Perspective. *Korean Studies*, vol. 28, pp. 14–48, 2004.

Bar-Oz, G., Nahshoni, P., Motro, H. and Oren, E. D. Symbolic Metal Bit and Saddlebag Fastenings in a Middle Bronze Age Donkey Burial. *PloS One*, vol. 8, no. 3, p. e58648, March 6, 2013.

Barrón-Ortiz, C. I., Avilla, L. S., Jass, C. N., Bravo-Cuevas, V. M., Machado, H. and Mothé, D. What Is Equus? Reconciling Taxonomy and Phylogenetic Analyses. *Frontiers in Ecology and Evolution*, vol. 7, p. 343, 2019.

Barros Damgaard, P. de, Martiniano, R., Kamm, J., Moreno-Mayar, J. V., Kroonen, G., Peyrot, M., Barjamovic, G., et al. The First Horse Herders and the Impact of Early Bronze Age Steppe Expansions into Asia. *Science*, vol. 360, no. 6396. https://doi.org/10.1126/science.aar7711.

Baumer, C. *The History of Central Asia*. Vol. 1, *The Age of the Steppe Warriors*. I. B. Tauris, 2012.

Bayarsaikhan, J. Монголын умард нутгийн буган хөшөөдд [Deer Stones of Northern Mongolia]. National Museum of Mongolia, 2017.

Bayarsaikhan, J., Turbat, T., Bayandelger, C., Tuvshinjargal, T., Wang, J., Chechushkov, I., Uetsuki, M., et al. The Origins of Saddles and Riding Technology in East Asia: New Discoveries from the Mongolian Altai. *Antiquity*, vol. 98, no. 397, pp. 102–18, 2024. https://doi.org/10.15184/aqy.2023.172.

Beck, U., Wagner, M., Li, X., Durkin-Meisterernst, D. and Tarasov, P. E. The Invention of Trousers and Its Likely Affiliation with Horseback Riding and Mobility: A Case Study of Late 2nd Millennium BC Finds from Turfan in Eastern Central Asia. *Quaternary International*, vol. 348, pp. 224–35, October 20, 2014.

Beckwith, C. I. *Empires of the Silk Road*. Princeton University Press, 2009.

Belardi, J. B., Marina, F. C., Madrid, P., Barrientos, G. and Campan, P. Late Holocene Guanaco Hunting Grounds in Southern Patagonia: Blinds, Tactics and Differential Landscape Use. *Antiquity*, vol. 91, no. 357, pp. 718–31, June 2017.

Bendrey, R. An Analysis of Factors Affecting the Development of an Equid Cranial Enthesopathy. *Veterinarija Ir Zootechnika*, vol. 41, no. 63, pp. 25–31, 2008.

Bendrey, R. From Wild Horses to Domestic Horses: A European Perspective. *World Archaeology*, vol. 44, no. 1, pp. 135–57, March 1, 2012.

Bendrey, R. The Horse. In *Extinctions and Invasions: A Social History of British Fauna*, T. O'Connor and N. J. Sykes, Eds., pp. 10–16. Windgather Press, 2010.

Bendrey, R. New Methods for the Identification of Evidence for Bitting on Horse Remains from Archaeological Sites. *Journal of Archaeological Science*, vol. 34, no. 7, pp. 1036–50, July 1, 2007.

Bendrey, R., Hayes, T. E. and Palmer, M. R. Patterns of Iron Age Horse Supply: An Analysis of Strontium Isotope Ratios in Teeth. *Archaeometry*, vol. 51, no. 1, pp. 140–50, January 2009.

Benecke, N. The Horse Skeletons from the Scythian Royal Grave Mound at Arzan 2 (Tuva, W. Siberia). *Documenta Archaeobiologiae*, vol. 5, pp. 115–31, 2007.

Benecke, N. Late Prehistoric Exploitation of Horses in Central Germany and Neighboring Areas—The Archaeozoological Record. In *Horses and Humans: The Evolution of Human-Equine Relationships*, S. Olsen, S. Grant, A. Choyke, and L. Bartosiewicz, Eds., pp. 195–208. British Archaeological Reports, 2006.

Benecke, N. and Driesch, A. von den. Horse Exploitation in the Kazakh Steppes during the Eneolithic and Bronze Age. In *Prehistoric Steppe Adaptation and the Horse*, M. Levine, C. Renfrew, and Katie Boyle, Eds., pp. 69–82. McDonald Institute for Archaeological Research, 2003.

Bennett, D. *Conquerors: The Roots of New World Horsemanship*. Amigo, 1998.

Bennett, E. A., Champlot, S., Peters, J., Arbuckle, B. S., Guimaraes, S., Pruvost, M., Bar-David, S., et al. Taming the Late Quaternary Phylogeography of the Eurasiatic Wild Ass through Ancient and Modern DNA. *PloS One*, vol. 12, no. 4, p. e0174216, April 19, 2017.

Bennett, E. A., Weber, J., Bendhafer, W., Champlot, S., Peters, J., Schwartz, G. M., Grange, T. and Geigl, E.-M. The Genetic Identity of the Earliest

Human-Made Hybrid Animals, the Kungas of Syro-Mesopotamia. *Science Advances*, vol. 8, no. 2, p. eabm0218, January 14, 2022.

Bento, C. M. and Pereira, P. V. R. *History of the Azores*. EGA Rua Manuel Augusto de Amaral, 1994.

Bergin, B. and Bergin, B. *The Hawaiian Horse*. University of Hawai'i Press, 2017.

Bergström, A., Stanton, D. W. G., Taron, U. H., Frantz, L., Sinding, M.-H. S., Ersmark, E., Pfrengle, S., et al. Grey Wolf Genomic History Reveals a Dual Ancestry of Dogs. *Nature*, June 29, 2022. https://doi.org/10.1038/s41586-022-04824-9.

Bernor, R. L., Cirilli, O., Jukar, A. M., Potts, R., Buskianidze, M. and Rook, L. Evolution of Early Equus in Italy, Georgia, the Indian Subcontinent, East Africa, and the Origins of African Zebras. *Frontiers in Ecology and Evolution*, vol. 7, 2019. https://doi.org/10.3389/fevo.2019.00166.

Bessac, F. B., Bessac, S. L. and Steelquist, J. O. B. *Death on the Chang Tang: Tibet, 1950: The Education of an Anthropologist*. University of Montana Press, 2006.

Biknevicius, A. R., Mullineaux, D. R. and Clayton, H. M. Locomotor Mechanics of the Tölt in Icelandic Horses. *American Journal of Veterinary Research*, vol. 67, no. 9, pp. 1505-10, September 2006.

Bjørn, R. Indo-European Loanwords and Exchange in Bronze Age Central and East Asia. *Evolutionary Human Sciences*, vol. 4, 2022. https://doi.org/10.1017/ehs.2022.16.

Blench, R. The Austronesians in Madagascar and Their Interaction with the Bantu of the East African Coast: Surveying the Linguistic Evidence for Domestic and Translocated Animals. *Studies in Philippine Languages and Cultures*, vol. 18, pp. 18-43, 2008.

Blench, R. Wild Asses and Donkeys in Africa: Interdisciplinary Evidence for Their Biogeography, History and Current Use. *Proceedings of the 9th Donkey Conference, School of Oriental and African Studies, London, UK*, pp. 8-9, 2012.

Bliujienė, A. and Butkus, D. Burials with Horses and Equestrian Equipment on the Lithuanian and Latvian Littorals and Hinterlands (from the Fifth to the Eighth Centuries). *Archaeologia Baltica*, vol. 11, pp. 149-63, 2009.

Boeskorov, G. G., Potapova, O. R., Protopopov, A. V., Plotnikov, V. V., Maschenko, E. N., Shchelchkova, M. V., Petrova, E. A., et al. A Study of a

Frozen Mummy of a Wild Horse from the Holocene of Yakutia, East Siberia, Russia. *Mammal Research*, vol. 63, no. 3, pp. 307–14, July 2018.

Bökönyi, S. and Bartosiewicz, L. A Review of Animal Remains from Shahr-I Sokhta (Eastern Iran). In *Archaeozoology of the Near East IV B*, M. Mashkour, A. M. Choyke, H. Buitenhuis, and F. Poplin, Eds., pp. 116–52. ARC, 2000.

Borrero, L. A. and Martin, F. M. Fragmented Records: Fuego-Patagonian Hunter-Gatherers and Archaeological Change. In *Archaeology on the Threshold: Studies in the Processes of Change*, J. Wardle, R. Hitchcock, M. Schmader, and P. Yu, Eds., pp. 68–88, University Press of Florida, 2023.

Bourgeon, L., Burke, A. and Higham, T. Earliest Human Presence in North America Dated to the Last Glacial Maximum: New Radiocarbon Dates from Bluefish Caves, Canada. *PloS One*, vol. 12, no. 1, p. e0169486, January 6, 2017.

Bourne, B. F. *The Giants of Patagonia: Captain Bourne's Account of His Captivity amongst the Extra-Ordinary Savages of Patagonia*. Ingram, Cooke, 1853.

Brink, J. W., Barrón-Ortiz, C. I., Loftis, K. and Speakmart, R. J. Pleistocene Horse and Possible Human Association in Central Alberta, 12,700 Years Ago. *Canadian Journal of Archaeology*, vol. 41, no. 1, pp. 79–96, 2017.

Brown, D. and Anthony, D. Bit Wear, Horseback Riding and the Botai Site in Kazakstan. *Journal of Archaeological Science*, vol. 25, no. 4, pp. 331–47, 1998.

Brubaker, L. and Udell, M. A. R. Cognition and Learning in Horses (*Equus caballus*): What We Know and Why We Should Ask More. *Behavioural Processes*, vol. 126, pp. 121–31, May 2016.

Burke, A. and Castanet, J. Histological Observations of Cementum Growth in Horse Teeth and Their Application to Archaeology. *Journal of Archaeological Science*, vol. 22, no. 4, pp. 479–93, 1995.

Burmeister, S., Krispijn, T. J. H. and Raulwing, P. Some Notes on Pictograms Interpreted as Sledges and Wheeled Vehicles in the Archaic Texts from Uruk. In *Equids and Wheeled Vehicles in the Ancient World: Essays in Memory of Mary A. Littauer*, BAR International Series 2923, P. Raulwing, K. M. Linduff, and J. Crouwel, Eds., pp. 49–70. British Archaeological Reports, 2019.

Cabrera, A. *Caballos de América*. Editorial Sudamericana, 1945.

Cai, D., Zhu, S., Gong, M., Zhang, N., Wen, J., Liang, Q., Sun, W., et al. Radiocarbon and Genomic Evidence for the Survival of Equus

Sussemionus until the Late Holocene. *eLife*, vol. 11, May 11, 2022. https://doi.org/10.7554/eLife.73346.

Camphora, A. L. *Animals and Society in Brazil: From the Sixteenth to Nineteenth Centuries*. White Horse, 2021.

Caprioli, M. Equestrian Military Equipment of the Eastern Roman Armies in the Sixth and Seventh Centuries. In *The Materiality of the Horse*, M. Bibby and B. Scott, Eds., pp. 221–37. Trivent, 2020.

Carlson, R. L. *Eighteenth Century Navajo Fortresses of the Gobernador District*. University of Colorado Press, 1965.

Carver, M. O. H. and Carver, M. *Sutton Hoo: Burial Ground of Kings?* University of Pennsylvania Press, 1998.

Caspari, G., Sadykov, T., Blochin, J. and Hajdas, I. Tunnug 1 (Arzhan 0)—An Early Scythian Kurgan in Tuva Republic, Russia. *Archaeological Research in Asia*, vol. 15, pp. 82–87, November 1, 2017.

Catlin, G., *North American Indians*. Penguin, 2004.

Chan, M. B., Ed. *Nomadic Empires of the Mongolian Steppes*. National Museum of Korea, 2018.

Chard, T. Did the First Spanish Horses Landed in Florida and Carolina Leave Progeny? *American Anthropologist*, vol. 42, no. 1, pp. 90–106, 1940.

Chechushkov, I. V. and Epimakhov, A. V. Eurasian Steppe Chariots and Social Complexity during the Bronze Age. *Journal of World Prehistory*, vol. 31, no. 4, pp. 435–83, December 1, 2018.

Chechushkov, I. V. and Epimakhov, A. V. Relative and Absolute Chronologies of the Chariot Complex in Northern Eurasia and Early Indo-European Migrations. In *The Indo-European Puzzle Revisited: Integrating Archaeology, Genetics, and Linguistics*, K. Kristiansen, G. Kroonen, and E. Willerslev, Eds., pp. 501–20. Cambridge University Press, 2023.

Chechushkov, I. V., Usmanova, E. R. and Kosintsev, P. A. Early Evidence for Horse Utilization in the Eurasian Steppes and the Case of the Novoil'inovskiy 2 Cemetery in Kazakhstan. *Journal of Archaeological Science, Reports*, vol. 32, no. 102420, p. 102420, August 1, 2020.

Chiga, H. 日本に伝えられた馬文化 [Horse Culture Passed Down to Japan]. In 馬の考古学 [Archaeology of Horses], K. Migishima, Ed., pp. 12–21. Yuzankaku, 2019.

Clarence-Smith, W. G. Breeding and Power in Southeast Asia: Horses, Mules and Donkeys in the Longue Durée. In *Environment, Trade, and*

Society in Southeast Asia, D. Henley and H.S. Nordholt, Eds., pp. 32-45. Brill, 2015.

Clarence-Smith, W.G. Elephants, Horses, and the Coming of Islam to Northern Sumatra. *Indonesia and the Malay World*, vol. 32, no. 93, pp. 271-84, July 1, 2004.

Clements, N.P. *The Black War: Fear, Sex and Resistance in Tasmania*. University of Queensland Press, 2014.

Clutton-Brock, J. The Buhen Horse. *Journal of Archaeological Science*, vol. 1, no. 1, pp. 89-100, March 1, 1974.

Clutton-Brock, J. Were the Donkeys at Tell Brak (Syria) Harnessed with a Bit? In *Prehistoric Steppe Adaptation and the Horse*, M. Levine, C. Renfrew, and K. Boyle, Eds., pp. 126-28. McDonald Institute for Archaeological Research, 2003.

Clutton-Brock, J. and Davies, S. More Donkeys from Tell Brak. *Iraq*, vol. 55, pp. 209-21, October 1993.

Collin, Y.R.H. The Relationship between the Indigenous Peoples of the Americas and the Horse: Deconstructing a Eurocentric Myth. PhD diss., University of Alaska Fairbanks, 2017.

Cook, W.R. Damage by the Bit to the Equine Interdental Space and Second Lower Premolar. *Equine Veterinary Education*, vol. 23, no. 7, pp. 355-60, July 18, 2011.

Cooke, B. *Imperial China: The Art of the Horse in Chinese History: Exhibition Catalog*. Art Media Resources, 2000.

Cooper, H.K., Mason, O.K., Mair, V., Hoffecker, J.F. and Speakman, R.J. Evidence of Eurasian Metal Alloys on the Alaskan Coast in Prehistory. *Journal of Archaeological Science*, vol. 74, pp. 176-83, October 1, 2016.

Corbino, C.A., Comegna, C., Amoretti, V., Modi, A., Cannariato, C., Lari, M., Caramelli, D. and Osanna, M. Equine Exploitation at Pompeii (AD 79). *Journal of Archaeological Science: Reports*, vol. 48, 2023.

Cowdrey, M., Martin, N. and Coleman, W. *American Indian Horse Masks*. Hawk Hill, 2006.

Creel, H.G. The Role of the Horse in Chinese History. *American Historical Review*, vol. 70, no. 3, pp. 647-72, 1965.

Crisman, K.J. and Cohn, A.B. *When Horses Walked on Water*. Smithsonian, 1998.

Crosby, A.W. *Ecological Imperialism: The Biological Expansion of Europe, 900-1900*. Cambridge University Press, 2004.

Crouwel, J. H. *Chariots and Other Wheeled Vehicles in Iron Age Greece.* Allard Pierson Museum, 1992.

Cultural Relics Team of Xinjiang Museum. 且末县扎滚鲁克五座墓葬发掘简报 [Excavation Report on Five Burials at the Zaghunluq Cemetery in Qiemo County]. 新疆文物 [Xinjiang Wenwu], vol. 3, pp. 2–18, 1998.

Curta, F. The Earliest Avar-Age Stirrups, or the "Stirrup Controversy" Revisited. In *The Other Europe in the Middle Ages*, pp. 297–326. Brill, 2008.

Czebreszuk, J., Kośko, A. and Szmyt, M. The Horse, Wagon, and Roads. In *Origin and Spreading of Chariots*, A. I. Vasilenko, Ed., pp. 47–54. Globus, 2008.

Daly, K. G., Maisano Delser, P., Mullin, V. E., Scheu, A., Mattiangeli, V., Teasdale, M. D., Hare, A. J., et al. Ancient Goat Genomes Reveal Mosaic Domestication in the Fertile Crescent. *Science*, vol. 361, no. 6397, pp. 85–88, July 6, 2018.

D'Amato, R. and Salimbeti, A. *Sea Peoples of the Bronze Age Mediterranean c. 1400 BC–1000 BC.* Bloomsbury USA, 2015.

Danino, M. The Horse and the Aryan Debate. *Journal of Indian History and Culture*, vol. 13, pp. 33–39, 2006.

Darwin, C. *Journal of Researches into the Natural History and Geology of the Countries Visited during the Voyage of H. M. S. "Beagle" Round the World: Under the Command of Capt. Fitz Roy, R. N.* Ward, Lock and Company, 1889.

Davis, L. G., Madsen, D. B., Becerra-Valdivia, L., Higham, T., Sisson, D. A., Skinner, S. M., Stueber, D., et al. Late Upper Paleolithic Occupation at Cooper's Ferry, Idaho, USA, ~16,000 Years Ago. *Science*, vol. 365, no. 6456, pp. 891–97, August 30, 2019.

Deagan, K. A. and Cruxent, J. M. *Columbus's Outpost among the Taínos: Spain and America at La Isabela, 1493–1498.* Yale University Press, 2008.

Dean, B. A Descriptive Label for Spurs. *Metropolitan Museum of Art Bulletin*, vol. 11, no. 10, pp. 217–19, 1916.

deFrance, S. D. Diet and Animal Use in Colonial Potosi. *Chungara: Revista de Antropologia Chilena*, vol. 44, pp. 9–24, 2012.

deFrance, S. D. Iberian Foodways in the Moquegua and Torata Valleys of Southern Peru. *Historical Archaeology*, vol. 30, no. 3, pp. 20–48, September 1, 1996.

de Lacy, C. *The History of the Spur*. The Connoisseur (Otto Limited), 1911.

Delaunay, A. N., Belardi, J. B., Marina, F. C., Saletta, M. J. and De Angelis, H. Glass and Stoneware Knapped Tools among Hunter-Gatherers in Southern Patagonia and Tierra del Fuego. *Antiquity*, vol. 91, no. 359, pp. 1330–43, October 2017.

Delsol, N., Stucky, B. J., Oswald, J. A., Reitz, E. J., Emery, K. F. and Guralnick, R. Analysis of the Earliest Complete mtDNA Genome of a Caribbean Colonial Horse (*Equus caballus*) from 16th-Century Haiti. *PloS One*, vol. 17, no. 7, p. e0270600, July 27, 2022.

Deng, T. The Fossils of the Przewalski's Horse and the Climatic Variations of the Late Pleistocene in China. In *Equids in Time and Space: Papers in Honour of Vera Eisenmann*, M. Mashkour, Ed., pp. 12–19. Oxbow, 2006.

Dennis, S., Meyers, A. and Mitchell, P. Diseases and Distributions of Wild and Domestic Equids. In *The Equids: A Suite of Splendid Species*, H. H. T. Prins and I. J. Gordon, Eds., pp. 269–98. Springer, 2023.

Der Sarkissian, C., Ermini, L., Schubert, M., Yang, M. A., Librado, P., Fumagalli, M., Jónsson, H., et al. Evolutionary Genomics and Conservation of the Endangered Przewalski's Horse. *Current Biology*, vol. 25, no. 19, pp. 2577–83, October 5, 2015.

Diakakis, N. and Tyrnenopoulou, P. Correlation between Equine Colic and Weather Changes. *Journal of the Hellenic Veterinary Medical Society*, vol. 68, no, 3, pp. 455–66, 2018.

Dien, A. E. The Stirrup and Its Effect on Chinese Military History. *Ars Orientalis*, vol. 16, pp. 33–56, 1986.

Dietz, U. L. Horseback Riding: Man's Access to Speed. In *Prehistoric Steppe Adaptation and the Horse*, M. Levine, C. Renfrew, and K. Boyle, Eds., pp. 189–99. McDonald Institute for Archaeological Research, 2003.

Dixon, E. J. *Bones, Boats and Bison: Archeology and the First Colonization of Western North America*. University of New Mexico Press, 1999.

Dobat, A. S., Douglas Price, T., Kveiborg, J., Ilkjær, J. and Rowley-Conwy, P. The Four Horses of an Iron Age Apocalypse: War-Horses from the Third-Century Weapon Sacrifice at Illerup Aadal (Denmark). *Antiquity*, vol. 88, no. 339, pp. 191–204, March 2014.

Doxey, D. Napatan Horses and the Horse Cemetery at El-Kurru, Sudan. In *Equids and Wheeled Vehicles in the Ancient World: Essays in Memory of Mary*

A. Littauer, P. Raulwing, K. M. Linduff, and J. H. Crouwel, Eds., pp. 137–48. BAR, 2019.

Drass, R. R., Perkins, S. M. and Vehik, S. C. Digging Ditches: Archaeological Investigations of Historically Reported Fortifications at Bryson-Paddock (34KA5) and Other Southern Plains Village Sites. In *Archaeological Perspectives on Warfare on the Great Plains*, A. Clark and D. Bamforth, Eds., pp. 211–36. University of Colorado Press, 2018.

Drews, R. *The Coming of the Greeks: Indo-European Conquests in the Aegean and the Near East*. Princeton University Press, 1994.

Drews, R. *Early Riders: The Beginnings of Mounted Warfare in Asia and Europe*. Routledge, 2004.

Dudd, S. N., Evershed, R. P. and Levine, M. Organic Residue Analysis of Lipids in Potsherds from the Early Neolithic Settlement of Botai, Kazakhstan. In *Prehistoric Steppe Adaptation and the Horse*, M. Levine, C. Renfrew, and K. Boyle, Eds., pp. 45–53. McDonald Institute for Archaeological Research, 2003.

Enghoff, I. B. *Hunting, Fishing and Animal Husbandry at the Farm Beneath the Sand, Western Greenland*. Museum Tusculanum Press, 2003.

Erdenebaatar, D. Material Cultural Heritage of the Xiongnu Empire. Известия Лаборатории древних технологий [Journal of Ancient Technology Laboratory], vol. 14, no. 2 (27), pp. 54–73, 2018.

Erdene-Ochir, N., Tseveendorj, D., Polosmak, N. V. and Bogdanov, E. S. Ноён уулын дурсгалын археологийн шинэ судалгаа [New Archaeological Research at the Site of Noyon Uul]. Mongolian Academy of Sciences, 2021.

Eregzen, G. *Ancient Funeral Monuments of Mongolia*. Mongolian Academy of Sciences, Ulaanbaatar, 2016.

Esin, Y. N. An Eastern Tibetan Tool for Managing Draught Cattle. *Archaeology, Ethnology & Anthropology of Eurasia*, vol. 48, no. 3, pp. 107–16, 2020.

Esin, Y. N., Magail, J., Gantulga, J.-O. and Yeruul-Erdene, C. Chariots in the Bronze Age of Central Mongolia Based on the Materials from the Khoid Tamir River Valley. *Archaeological Research in Asia*, vol. 27, p. 100304, September 1, 2021.

Esin, Y. N., Magail, J. and Yeruul-Erdene, C. Paint on Deer Stones of Mongolia. *Archaeology, Ethnology, & Anthropology of Eurasia*, vol. 45, no. 3, pp. 79–89, 2017.

Ewers, J. C. *The Horse in Blackfoot Indian Culture: With Comparative Material from Other Western Tribes.* Literary Licensing, 2011. First published 1955 by US Government Printing Office (Washington, DC).

Fages, A., Hanghøj, K., Khan, N., Gaunitz, C., Seguin-Orlando, A., Leonardi, M., McCrory Constantz, C., et al. Tracking Five Millennia of Horse Management with Extensive Ancient Genome Time Series. *Cell*, vol. 177, no. 6, pp. 1419-35.e31, May 30, 2019.

Fages, A., Seguin-Orlando, A., Germonpré, M. and Orlando, L. Horse Males Became Over-Represented in Archaeological Assemblages during the Bronze Age. *Journal of Archaeological Science: Reports*, vol. 31, p. 102364, June 1, 2020.

Faith, J. T. Late Pleistocene and Holocene Mammal Extinctions on Continental Africa. *Earth-Science Reviews*, vol. 128, pp. 105-21, January 1, 2014.

Fan, R., Gu, Z., Guang, X., Marín, J. C., Varas, V., González, B. A., Wheeler, J. C., et al. Genomic Analysis of the Domestication and Post-Spanish Conquest Evolution of the Llama and Alpaca. *Genome Biology*, vol. 21, no. 1, p. 159, July 2, 2020.

Fenner, J. N., Delgermaa, L., Piper, P. J., Wood, R. and Stuart-Williams, H. Stable Isotope and Radiocarbon Analyses of Livestock from the Mongol Empire Site of Avraga, Mongolia. *Archaeological Research in Asia*, vol. 22, p. 100181, June 1, 2020.

Fijn, N. Encountering the Horse: Initial Reactions of Aboriginal Australians to a Domesticated Animal. *Australian Humanities Review*, vol. 62, pp. 1-25, 2017.

Fillios, M. A. and Taçon, P. S. C. Who Let the Dogs In? A Review of the Recent Genetic Evidence for the Introduction of the Dingo to Australia and Implications for the Movement of People. *Journal of Archaeological Science: Reports*, vol. 7, pp. 782-92, 2016.

Fitzhugh, W. W. The Mongolian Deer Stone-Khirigsuur Complex: Dating and Organization of a Late Bronze Age Menagerie. In *Current Archaeological Research in Mongolia: Papers from the First International Conference on "Archaeological Research in Mongolia" Held in Ulaanbaatar, August 19-23rd, 2007*, J. Bemmann, H. Parzinger, E. Pohl, and D. Tseveendorzh, Eds., pp. 183-99. Rheinische Friedrich-Wilhelms-Universitat, 2009.

Forbes, C. *Australia on Horseback.* Macmillan Australia, 2014.

Forbes, J.D. The Appearance of the Mounted Indian in Northern Mexico and the Southwest, to 1680. *Southwestern Journal of Anthropology*, vol. 15, no. 2, pp. 189–212, July 1, 1959.

Foronova, I. Late Quaternary Equids (genus *Equus*) of South-Western and South-Central Siberia. In *Equids in Time and Space: Papers in Honour of Vera Eisenmann*, M. Mashkour, Ed., pp. 20–30. Oxbow, 2006.

French, C. and Kousoulakou, M. Geomorphological and Micromorphological Investigations of Palaeosols, Valley Sediments and a Sunken Floored Dwelling at Botai, Kazakhstan. In *Prehistoric Steppe Adaptation and the Horse*, M. Levine, C. Renfrew, and K. Boyle, Eds., pp. 105–14. McDonald Institute for Archaeological Research, 2003.

Gaastra, J.S., Greenfield, H.J. and Vander Linden, M. Gaining Traction on Cattle Exploitation: Zooarchaeological Evidence from the Neolithic Western Balkans. *Antiquity*, vol. 92, no. 366, pp. 1462–77, December 2018.

Gallardo, F., Castro, V. and Miranda, P. Riders on the Storm: Rock Art in the Atacama Desert (Northern Chile). *World Archaeology*, vol. 31, no. 2, pp. 225–42, October 1, 1999.

Gaunitz, C., Fages, A., Hanghøj, K., Albrechtsen, A., Khan, N., Schubert, M., Seguin-Orlando, A., et al. Ancient Genomes Revisit the Ancestry of Domestic and Przewalski's Horses. *Science*, February 22, 2018. https://doi.org/10.1126/science.aao3297.

Giardina, M., Otaola, C. and Franchetti, F. Hunting, Butchering and Consumption of Rheidae in the South of South America: An Actualistic Study. In *Ancient Hunting Strategies in Southern South America*, J.B. Belardi, D.L. Bozzuto, P.M. Fernández, E.A. Moreno, and G.A. Neme, Eds., pp. 159–74. Springer International, 2021.

Gifford-Gonzalez, D. and Sunseri, J. Foodways on the Frontier: Animal Use and Identity in Early Colonial New Mexico. In *The Archaeology of Food and Identity*, K.C. Twiss, Ed., pp. 260–87. Center for Archaeological Investigations, Southern Illinois University, 2007.

Gilbert, A.S. Equid Remains from Godin Tepe, Western Iran: An Interim Summary and Interpretation, with Notes on the Introduction of the Horse into Southwest Asia. *Equids in the Ancient World II. Beihefte Zum Tübinger Atlas des Vorderen Orients. Reihe A (Naturwissenchaften)*, vol. 19, no. 2, pp. 75–123, 1991.

Gimbutas, M. *The Kurgan Culture and the Indo-Europeanization of Europe: Selected Articles from 1952 to 1993*. Institute for the Study of Man, 1997.

Glover, I. *Archaeology in Eastern Timor, 1966–67*. Terra Australis, 1986.

Glover, I. Excavations in Timor: A Study of Economic Change and Cultural Continuity in Prehistory. PhD diss., Australian National University, 1972.

Gnecchi-Ruscone, G. A., Khussainova, E., Kahbatkyzy, N., Musralina, L., Spyrou, M. A., Bianco, R. A., Radzeviciute, R., et al. Ancient Genomic Time Transect from the Central Asian Steppe Unravels the History of the Scythians. *Science Advances*, vol. 7, no. 13, March 2021. https://doi.org /10.1126/sciadv.abe4414.

Goldhahn, J. Bredarör on Kivik: A Monumental Cairn and the History of Its Interpretation. *Antiquity*, vol. 83, no. 320, pp. 359–71, 2009.

Greenfield, H. J. The Social and Economic Context for Domestic Horse Origins in Southeastern Europe: A View from Ljuljaci in the Central Balkans. In *Horses and Humans: The Evolution of Human-Equine Relationships*, S. Olsen, S. Grant, A. Choyke, and L. Bartosiewicz, Eds., pp. 221–44. British Archaeological Reports, 2006.

Greenfield, H. J., Greenfield, T. L., Arnold, E., Shai, I., Albaz, S. and Maeir, A. M. Evidence for Movement of Goods and Animals from Egypt to Canaan during the Early Bronze of the Southern Levant: A View from Tell eş-Şâfi/Gath. *Ägypten und Levante*, vol. 30, pp. 377–97, 2020.

Greenfield, H. J., Shai, I., Greenfield, T. L., Arnold, E. R., Brown, A., Eliyahu, A. and Maeir, A. M. Earliest Evidence for Equid Bit Wear in the Ancient Near East: The "Ass" from Early Bronze Age Tell Eş-Şâfi/Gath, Israel. *PloS One*, vol. 13, no. 5, p. e0196335, May 16, 2018.

Griggs, J. C. *The Archaeology of Central Caribbean Panama*. University of Texas at Austin, 2005.

Grinnell, G. B., *Pawnee Hero Stories and Folk-Tales: With Notes on the Origin, Customs and Character of the Pawnee People*. D. Nutt, 1893.

Guedes, J. d'Alpoim and Aldenderfer, M. The Archaeology of the Early Tibetan Plateau: New Research on the Initial Peopling through the Early Bronze Age. *Journal of Archaeological Research*, vol. 28, no. 3, pp. 339–92, September 2020.

Guimaraes, S., Arbuckle, B. S., Peters, J., Adcock, S. E., Buitenhuis, H., Chazin, H., Manaseryan, N., et al. Ancient DNA Shows Domestic Horses Were Introduced in the Southern Caucasus and Anatolia during the Bronze

Age. *Science Advances*, vol. 6, no. 38, September 2020. https://doi.org/10.1126/sciadv.abb0030.

Guinnard, A. *Three Years' Slavery among the Patagonians: An Account of His Captivity*. R. Bentley, 1871.

Guthrie, R. D. Human-Horse Relations Using Paleolithic Art: Pleistocene Horses Drawn from Life. In *Horses and Humans: The Evolution of Human-Equine Relationships*, S. Olsen, S. Grant, A. Choyke, and L. Bartosiewicz, Eds., pp. 61–80. British Archaeological Reports, 2006.

Guthrie, R. D. Rapid Body Size Decline in Alaskan Pleistocene Horses before Extinction. *Nature*, vol. 426, no. 6963, pp. 169–71, November 13, 2003.

Haak, W., Lazaridis, I., Patterson, N., Rohland, N., Mallick, S., Llamas, B., Brandt, G., et al. Massive Migration from the Steppe Was a Source for Indo-European Languages in Europe. *Nature*, vol. 522, no. 7555, pp. 207–11, June 11, 2015.

Habu, J. Seafaring and the Development of Cultural Complexity in Northeast Asia: Evidence from the Japanese Archipelago. In *The Global Origins and Development of Seafaring*, A. Anderson, J. H. Barrett, and K. V. Boyle, Eds., pp. 159–89, McDonald Institute for Archaeological Research, 2010.

Haile, E. W. *Jamestown Narratives: Eyewitness Accounts of the Virginia Colony, the First Decade, 1607-1617*. RoundHouse, 1998.

Haile, J., Froese, D. G., Macphee, R. D. E., Roberts, R. G., Arnold, L. J., Reyes, A. V., Rasmussen, M., et al. Ancient DNA Reveals Late Survival of Mammoth and Horse in Interior Alaska. *Proceedings of the National Academy of Sciences of the United States of America*, vol. 106, no. 52, pp. 22352–57, December 29, 2009.

Haines, F. The Northward Spread of Horses among the Plains Indians. *American Anthropologist*, vol. 40, no. 3, pp. 429–37, 1938.

Hämäläinen, P. *The Comanche Empire*. Yale University Press, 2008.

Hämäläinen, P. *Lakota America: A New History of Indigenous Power*. Yale University Press, 2019.

Hamilakis, Y. A Footnote on the Archaeology of Power: Animal Bones from a Mycenaean Chamber Tomb at Galatas, NE Peloponnese. *Annual of the British School at Athens*, vol. 91, pp. 153–66, November 1996.

Harbeck, M., Seifert, L., Hänsch, S., Wagner, D. M., Birdsell, D., Parise, K. L., Wiechmann, I., et al. Yersinia Pestis DNA from Skeletal Remains from the

6th Century AD Reveals Insights into Justinianic Plague. *PLoS Pathogens*, vol. 9, no. 5, p. e1003349, May 2, 2013.

Harding, A. Horse-Harness and the Origins of the Mycenean Civilisation. In *Autochthon: Papers Presented to O. T. P. K. Dickinson*, A. Dakouri-Hild and S. Sherratt, Eds., pp. 296–300. British Archaeological Reports, 2005.

Hayhurst, Y. A Recent Find of a Horse Skull in a House at Ballaugh, Isle of Man. *Folklore*, vol. 100, no. 1, pp. 105–9, January 1, 1989.

Heintzman, P. D., Zazula, G. D., MacPhee, R. D. E., Scott, E., Cahill, J. A., McHorse, B. K., Kapp, J. D., et al. A New Genus of Horse from Pleistocene North America. *eLife*, vol. 6, November 28, 2017. https://doi.org/0.7554/eLife.29944.

Hermes, T. R., Tishkin, A. A., Kosintsev, P. A., Stepanova, N. F., Krause-Kyora, B. and Makarewicz, C. A. Mitochondrial DNA of Domesticated Sheep Confirms Pastoralist Component of Afanasievo Subsistence Economy in the Altai Mountains (3300–2900 Cal BC). *Archaeological Research in Asia*, vol. 24, p. 100232, December 1, 2020.

Hildebrand, M. The Mechanics of Horse Legs. *American Scientist*, vol. 75, no. 6, pp. 594–601, 1987.

Hoffecker, J. F., Holliday, V. T., Stepanchuk, V. N. and Lisitsyn, S. N. The Hunting of Horse and the Problem of the Aurignacian on the Central Plain of Eastern Europe. *Quaternary International*, vol. 492, pp. 53–63, 2018.

Hoh, E. and Mair, V. H. *The True History of Tea*. Thames & Hudson, 2009.

Holder, P. *The Hoe and the Horse on the Plains: A Study of Cultural Development among North American Indians*. University of Nebraska Press, 1974.

Honeychurch, W. *Inner Asia and the Spatial Politics of Empire: Archaeology, Mobility, and Culture Contact*. Springer, 2015.

Horvath, V. Захоронения коней в камере №31 Кургана Аржан-1 (новые данные о культурных связях в Евразийских Степях в VIII–начале VI. в. до н.э.) [Horse Burials in Chamber No. 31 of Kurgan Arzhan-1 (New Data on the Cultural Connections in the Eurasian Steppes in the 8th–Early 6th Centuries BCE)]. Теория и практика археологических исследований [Theory and Practice of Archaeological Research], vol. 31, no. 3, pp. 134–53, 2020.

Horwitz, L. K., Rosen, S. A. and Bocquentin, F. A Late Neolithic Equid Offering from the Mortuary-Cult Site of Ramat Saharonim in the Central Negev. *Journal of the Israel Prehistoric Society* vol. 41, pp. 71-81, 2011.

Houle, J.-L. Emergent Complexity on the Mongolian Steppe: Mobility, Territoriality, and the Development of Early Nomadic Polities. PhD diss., University of Pittsburgh, 2010.

Houle, J.-L., Seitsonen, O., Égüez, N., Broderick, L. G., García-Granero, J. J. and Bayarsaikhan, J. Resilient Herders: A Deeply Stratified Multiperiod Habitation Site in Northwestern Mongolia. *Archaeological Research in Asia*, vol. 30, p. 100371, June 1, 2022.

Hudson, C. *Knights of Spain, Warriors of the Sun: Hernando de Soto and the South's Ancient Chiefdoms*. University of Georgia Press, 1998.

Hukantaival, S. Horse Skulls and "Alder Horse": The Horse as a Depositional Sacrifice in Buildings. *Archaeologia BALTICA*, vol. 11, pp. 350–56, 2009.

Hyland, A. *Equus: The Horse in the Roman World*. BT Batsford, 1990.

Iderkhangai, T. and Batjargal, B. Хүннүгийн Лунчэн, Чанюйтин, Лунтин, хэмээх үгсийн тухай, Луут хот хэмээн бичигдэх болсон шалтгаан [About the Xiongnu Period Words Longcheng, Chanyuting, and Longting, the Reason It Was Written as Dragon City]. Археологи, Түүх, Угсаатан судлалын сэтгүүл [Journal of the Archaeology, History, and Ethnography], vol. 16, no. 15, pp. 185–99, 2020.

Ikram, S. and Iskander, N. *Catalogue General of Egyptian Antiquities in the Cairo Museum*. Supreme Council of Antiquities Press, 2002.

Ingold, T. *Hunters, Pastoralists and Ranchers: Reindeer Economies and Their Transformations*. Cambridge University Press, 1980.

Ingstad, H. and Ingstad, A. S. *The Viking Discovery of America: The Excavation of a Norse Settlement in L'Anse aux Meadows, Newfoundland*. Breakwater, 2000.

Izbitser, E. V. Колесница с тормозом или реконструкции без тормозов [The Chariot with the Brake, or Reconstructions off the Handle]. *Stratum Plus*, no. 2, pp. 187–94, 2010.

Jacobson-Tepfer, E. *The Hunter, the Stag, and the Mother of Animals: Image, Monument, and Landscape in Ancient North Asia*. Oxford University Press, 2015.

Janz, L., Cameron, A., Bukhchuluun, D., Odsuren, D. and Dubreuil, L. Expanding Frontier and Building the Sphere in Arid East Asia. *Quaternary International*, vol. 559, pp. 150–64, September 10, 2020.

Jawad, A. Faunal Remains from Kalako-ḍeray, Swāt (Mid-2nd Millennium B.C.). *East and West*, vol. 48, no. 3/4, pp. 265–90, 1998.

Jeong, C., Wang, K., Wilkin, S., Taylor, W.T.T., Miller, B.K., Bemmann, J.H., Stahl, R., et al. A Dynamic 6,000-Year Genetic History of Eurasia's Eastern Steppe. *Cell*, vol. 183, no. 4, pp. 890-904.e29, November 12, 2020.

Johnson, E. M. *Lubbock Lake: Late Quaternary Studies on the Southern High Plains*. Texas A&M University Press, 1987.

Jones, P. *Ochre and Rust: Artefacts and Encounters on Australian Frontiers*. Oxford University Press, 2019.

Jones, R. A. Wings across the Silk Road: The Art of the Flying Horse in Early China. In *Imperial Horizons of the Silk Roads: Archaeological Case Studies*, B. Franicevic and M. N. Pareja, Eds., pp. 151-76. Archaeopress, 2023.

Jones, R. L. The Old French-Canadian Horse: Its History in Canada and the United States. *Canadian Historical Review*, vol. 28, no. 2, pp. 125-55, June 1, 1947.

Jong, I. de, Serna, A., Mange, E. and Prates, L. Mortuary Rituals and the Suttee among Mapuche Chiefdoms of Pampa-Patagonia: The Double Human Burial of Chimpay (Argentina). *Latin American Antiquity*, vol. 31, no. 4, pp. 838-52, December 2020.

Josephus. *The Life: Against Apion*. Translated by H. St. J. Thackeray. Loeb Classical Library 106. Harvard University Press, 1926.

Kaliff, A. and Oestigaard, T. *The Great Indo-European Horse Sacrifice: 4000 Years of Cosmological Continuity from Sintashta and the Steppe to Scandinavian Skeid*. Uppsala University, 2020.

Kaniewski, D., Van Campo, E., Van Lerberghe, K., Boiy, T., Vansteenhuyse, K., Jans, G., Nys, K., et al. The Sea Peoples, from Cuneiform Tablets to Carbon Dating. *PloS One*, vol. 6, no. 6, p. e20232, June 8, 2011.

Karczewska, M., Karczewski, M. and Gręzak, A. The Role of Horse Burials in the Bogaczewo Culture. Key Studies of Paprotki Kolonia Site 1 Cemetery, Northeast Poland. *Archaeologia Baltica*, vol. 18, pp. 97-108, 2012.

Karstens, S., Littleton, J., Frohlich, B., Amgaluntugs, T., Pearlstein, K. and Hunt, D. A Palaeopathological Analysis of Skeletal Remains from Bronze Age Mongolia. *Homo: Internationale Zeitschrift fur die Vergleichende Forschung am Menschen*, vol. 69, no. 6, pp. 324-34, November 2018.

Kausrud, K. L., Begon, M., Ari, T. B., Viljugrein, H., Esper, J., Büntgen, U., Leirs, H., et al. Modeling the Epidemiological History of Plague in Central Asia: Palaeoclimatic Forcing on a Disease System over the Past Millennium. *BMC Biology*, vol. 8, p. 112, August 27, 2010.

Kawami, T. S. That Strange Equid from Susa. In *Equids and Wheeled Vehicles in the Ancient World: Essays in Memory of Mary A. Littauer*, P. Raulwing, K. M. Linduff, and J. H. Crouwel, Eds., pp. 97–105. British Archaeological Reports, 2019.

Kefena, E., Dessie, T., Han, J. L., Kurtu, M. Y., Rosenbom, S. and Beja-Pereira, A. Morphological Diversities and Ecozones of Ethiopian Horse Populations. *Animal Genetic Resources/Resources Génétiques animales/Recursos Genéticos Animales*, vol. 50, pp. 1–12, June 2012.

Kelekna, P. *The Horse in Human History*. Cambridge University Press, 2009.

Kelekna, P. Northern Africa: Equestrian Penetration of the Sahara and the Sahel and Its Impact on Adjacent Regions. In *Equids and Wheeled Vehicles in the Ancient World: Essays in Memory of Mary A. Littauer*, P. Raulwing, K. Linduff, and J. Crouwel, Eds., British Archaeological Reports, pp. 123–36, 2019.

Kennedy, S. A. and VanValkenburgh, P. Zooarchaeology and Changing Food Practices at Carrizales, Peru following the Spanish Invasion. *International Journal of Historical Archaeology*, vol. 20, no. 1, pp. 73–104, March 1, 2016.

Kenoyer, J. M. Cultures and Societies of the Indus Tradition. In *Historical Roots in the Making of 'the Aryan,'* R. Thapar, Ed., pp. 21–49. National Book Trust, 2006.

Kimura, B., Marshall, F. B., Chen, S., Rosenbom, S., Moehlman, P. D., Tuross, N., Sabin, R. C., et al. Ancient DNA from Nubian and Somali Wild Ass Provides Insights into Donkey Ancestry and Domestication. *Proceedings. Biological Sciences/The Royal Society*, vol. 278, no. 1702, pp. 50–57, January 7, 2011.

Klæsøe, I. S. *Viking Trade and Settlement in Continental Western Europe*. Museum Tusculanum Press, 2010.

Kohl, P. L. *The Making of Bronze Age Eurasia*. Cambridge University Press, 2007.

Kooyman, B., Hills, L. V., McNeil, P. and Tolman, S. Late Pleistocene Horse Hunting at the Wally's Beach Site (DhPg-8), Canada. *American Antiquity*, vol. 71, no. 1, pp. 101–21, January 2006.

Khorvat, V. Захоронения коней в камере №31 Кургана Аржан-1 (новые данные о культурных связях в Евразийских Степях в VIII–начале VI в. до н.э.) [Horse Burials in Chamber No. 31 of Kurgan Arzhan-1 (New Data on the Cultural Connections in the Eurasian Steppes in the 8th–Early 6th

Centuries BCE)]. Теория и практика археологических исследований [Theory and Practice of Archaeological Research], vol. 3, no. 31, 134–53, 2020.

Kosintsev, P. A. The Human-Horse Relationship on the European-Asian Border in the Neolithic and Early Iron Age. In *Horses and Humans: The Evolution of Human-Equine Relationships*, S. Olsen, S. Grant, A. Choyke, and L. Bartosiewicz, Eds., pp. 127–36. British Archaeological Reports, 2006.

Koungoulos, L. and Fillios, M. Hunting Dogs Down Under? On the Aboriginal Use of Tame Dingoes in Dietary Game Acquisition and Its Relevance to Australian Prehistory. *Journal of Anthropological Archaeology*, vol. 58, p. 101146, June 1, 2020.

Kovalev, A. A. and Erdenebaatar, D. Discovery of New Cultures of the Bronze Age in Mongolia according to the Data Obtained by the International Central Asian Archaeological Expedition. In *Current Archaeological Research in Mongolia: Papers from the First International Conference on "Archaeological Research in Mongolia" Held in Ulaanbaatar, August 19–23rd, 2007*, J. Bemmann, H. Parzinger, E. Pohl, and D. Tseveendorzh, Eds., pp. 149–70. Rheinische Friedrich-Wilhelms-Universitat, 2009.

Kovalev, A. A. and Erdenebaatar, D. *Earliest European in the Heart of Asia: The Chemurchek Cultural Phenomena*. Vol. 2. Book Antiqua, 2014.

Kradin, N. N. Stateless Empire: The Structure of the Xiongnu Nomadic Super-Complex Chiefdom. *Xiongnu Archaeology: Multidisciplinary Perspectives of the First Steppe Empire in Inner Asia*, vol. 5, pp. 77–96, 2011.

Krueger, K. Social Ecology of Horses. In *Ecology of Social Evolution*, J. Korb and J. Heinze, Eds., pp. 195–206. Springer Berlin Heidelberg, 2008.

Kuitems, M., Wallace, B. L., Lindsay, C., Scifo, A., Doeve, P., Jenkins, K., Lindauer, S., et al. Evidence for European Presence in the Americas in AD 1021. *Nature*, October 20, 2021. https://doi.org/10.1038/s41586-021-03972-8.

Kumar, V. A Note on Chariot Burials Found at Sinauli District Baghpat UP. *Journal of Indian Ocean Archaeology*, vol. 3, no. 2, pp. 735–55, 2018.

Kuz'mina, E. E. Mythological Treatment of the Horse in Indo-European Culture. In *Horses and Humans: The Evolution of Human-Equine Relationships*, S. Olsen, S. Grant, A. Choyke, and L. Bartosiewicz, Eds., pp. 263–70. British Archaeological Reports, 2006.

Kuz'mina, E. E. *The Origin of the Indo-Iranians*. Brill, 2007.

Kuznetsov, P. F. The Emergence of Bronze Age Chariots in Eastern Europe. *Antiquity*, vol. 80, no. 309, pp. 638-45, September 2006.

Kyselý, R. and Peške, L. Horse Size and Domestication: Early Equid Bones from the Czech Republic in the European Context. *Anthropozoologica*, vol. 51, no. 1, pp. 15-39, June 2016.

Lahtinen, M., Clinnick, D., Mannermaa, K., Salonen, J. S. and Viranta, S. Excess Protein Enabled Dog Domestication during Severe Ice Age Winters. *Scientific Reports*, vol. 11, no. 1, p. 7, January 7, 2021.

Law, R. *The Horse in West African History: The Role of the Horse in the Societies of Pre-Colonial West Africa*. Routledge, 2018.

Lazaridis, I., Mittnik, A., Patterson, N., Mallick, S., Rohland, N., Pfrengle, S., Furtwängler, A., et al. Genetic Origins of the Minoans and Mycenaeans. *Nature*, vol. 548, no. 7666, pp. 214-18, August 2, 2017.

Lazzerini, N., Zazzo, A., Coulon, A., Marchina, C., Bayarkhuu, N., Bernard, V., Cervel, M., et al. Season of Death of Domestic Horses Deposited in a Ritual Complex from Bronze Age Mongolia: Insights from Oxygen Isotope Time-Series in Tooth Enamel. *Journal of Archaeological Science: Reports*, vol. 32, p. 102387, August 1, 2020.

Legrand, S. The Emergence of the Scythians: Bronze Age to Iron Age in South Siberia. *Antiquity*, vol. 80, no. 310, pp. 843-59, 2006.

Lepetz, S. Horse Sacrifice in a Pazyryk Culture Kurgan: The Princely Tomb of Berel' (Kazakhstan). Selection Criteria and Slaughter Procedures. *Anthropozoologica*, vol. 48, no. 2, pp. 309-21, December 1, 2013.

Lepetz, S., Clavel, B., Alioğlu, D., Chauvey, L., Schiavinato, S., Tonasso-Calvière, L., Liu, X., et al. Historical Management of Equine Resources in France from the Iron Age to the Modern Period. *Journal of Archaeological Science: Reports*, vol. 40, part B, 2021.

Lepetz, S., Debue, K. and Batsukh, D. To Accompany and Honour the Deceased. In *Masters of the Steppe: The Impact of the Scythians and Later Nomad Societies of Eurasia*, S. Pankova and S. Simpson, Eds., pp. 227-47. Archaeopress, 2021.

Lepetz, S., Zazzo, A., Bernard, V., Larminat, S. de, Magail, J. and Gantulga, J.-O. Customs, Rites, and Sacrifices Relating to a Mortuary Complex in Late Bronze Age Mongolia (Tsatsyn Ereg, Arkhangai). *Anthropozoologica*, vol. 54, no. 1, p. 151, November 8, 2019.

Letnic, M., Fillios, M. and Crowther, M. S. Could Direct Killing by Larger Dingoes Have Caused the Extinction of the Thylacine from Mainland Australia? *PloS One*, vol. 7, no. 5, e34877, 2012.

Levine, M. A. Chinese Chariot Horses and the Evolution of Horse Husbandry. McDonald Institute for Archaeological Research. https://www.arch.cam .ac.uk/~ml12/ChinPalaeoWebsite/introduction.htm.

Levine, M. A. Exploring the Criteria for Early Horse Domestication. In *Traces of Ancestry: Studies in Honor of Colin Renfrew*, M. Jones, Ed., pp. 115-26. McDonald Institute for Archaeological Research, 2004.

Levine, M. A. The Origins of Horse Husbandry on the Eurasian Steppe. In *Late Prehistoric Exploitation of the Eurasian Steppe*, pp. 5-58. McDonald Institute of Archaeological Research 1999.

Levine, M. A., Whitwell, K. E. and Jeffcott, L. B. Abnormal Thoracic Vertebrae and the Evolution of Horse Husbandry. *Archaeofauna*, vol. 14, pp. 93-109, 2005.

Li, F., Vanwezer, N., Boivin, N., Gao, X., Ott, F., Petraglia, M. and Roberts, P. Heading North: Late Pleistocene Environments and Human Dispersals in Central and Eastern Asia. *PloS One*, vol. 14, no. 5, p. e0216433, May 29, 2019.

Li, Y., Wu, L., Zhang, C., Liu, H., Huang, Z., Han, Y. and Yuan, J. Horses in Qin Mortuary Practice: New Insights from Emperor Qin Shihuang's Mausoleum. *Antiquity*, vol. 96, no. 388, pp. 903-19, 2022.

Li, Y., Zhang, C., Taylor, W. T. T., Chen, L., Flad, R. K., Boivin, N., Liu, H., et al. Early Evidence for Mounted Horseback Riding in Northwest China. *Proceedings of the National Academy of Sciences of the United States of America*, vol. 117, no. 47, pp. 29569-76, November 24, 2020.

Librado, P., Der Sarkissian, C., Ermini, L., Schubert, M., Jónsson, H., Albrechtsen, A., Fumagalli, M., et al. Tracking the Origins of Yakutian Horses and the Genetic Basis for Their Fast Adaptation to Subarctic Environments. *Proceedings of the National Academy of Sciences of the United States of America*, vol. 112, no. 50, pp. e6889-97, December 15, 2015.

Librado, P., Gamba, C., Gaunitz, C., Der Sarkissian, C., Pruvost, M., Albrechtsen, A., Fages, A., et al. Ancient Genomic Changes Associated with Domestication of the Horse. *Science*, vol. 356, no. 6336, pp. 442-45, April 28, 2017.

Librado, P., Khan, N., Fages, A., Kusliy, M. A., Suchan, T., Tonasso-Calvière, L., Schiavinato, S., et al. The Origins and Spread of Domestic Horses from the Western Eurasian Steppes. *Nature*, vol. 598, no. 7882, pp. 634–40, October 2021.

Lindner, S. Chariots in the Eurasian Steppe: A Bayesian Approach to the Emergence of Horse-Drawn Transport in the Early Second Millennium BC. *Antiquity*, vol. 94, no. 374, pp. 361–80, April 2020.

Littauer, M. A. Bits and Pieces. *Antiquity*, vol. 43, no. 172, pp. 289–300, December 1969.

Littauer, M. A. and Crouwel, J. H. *Chariots and Related Equipment from the Tomb of Tut'ankhamūn*. Griffith Institute, 1985.

Littauer, M. A. and Crouwel, J. H. *Wheeled Vehicles and Ridden Animals in the Ancient Near East*. Brill, 1979.

Littauer, M. A., Crouwel, J. H. and Raulwing, P. *Selected Writings on Chariots and Other Early Vehicles, Riding and Harness*. Brill, 2002.

Liu, X. Migration and Settlement of the Yuezhi-Kushan: Interaction and Interdependence of Nomadic and Sedentary Societies. *Journal of World History*, vol. 12, no. 2, pp. 261–92, 2001.

Liu, X., Zhang, Y., Liu, W., Li, Y., Pan, J., Pu, Y., Han, J., Orlando, L., Ma, Y. and Jiang, L. A Single-Nucleotide Mutation within the TBX3 Enhancer Increased Body Size in Chinese Horses. *Current Biology*, vol. 32, no. 2, pp. 480–87, 2021. https://doi.org/10.1016/j.cub.2021.11.052.

Löffelmann, T., Snoeck, C., Richards, J. D., Johnson, L. J., Claeys, P. and Montgomery, J. Sr Analyses from Only Known Scandinavian Cremation Cemetery in Britain Illuminate Early Viking Journey with Horse and Dog across the North Sea. *PloS One*, vol. 18, no. 2, p. e0280589, February 1, 2023.

Losey, R. J., Waters-Rist, A. L., Nomokonova, T. and Kharinskii, A. A. A Second Mortuary Hiatus on Lake Baikal in Siberia and the Arrival of Small-Scale Pastoralism. *Scientific Reports*, vol. 7, no. 1, p. 2319, May 24, 2017.

Lu, H., Zhang, J., Yang, Y., Yang, X., Xu, B., Yang, W., Tong, T., et al. Earliest Tea as Evidence for One Branch of the Silk Road across the Tibetan Plateau. *Scientific Reports*, vol. 6, p. 18955, January 7, 2016.

Macdonald, M. C. A. Hunting, Fighting, and Raiding: The Horse in Pre-Islamic Arabia. In *Furusiyya: The Horse in the Art of the Near East*, pp. 73–83. King Abdul Aziz Public Library, Riyadh, 2010.

MacEachern, S., Bourges, C. and Reeves, M. Early Horse Remains from Northern Cameroon. *Antiquity*, vol. 75, no. 287, pp. 62–67, 2001.

MacFadden, B. J. Fossil Horses—Evidence for Evolution. *Science*, vol. 307, no. 5716, pp. 1728-30, 2005.

MacFadden, B. J. *Fossil Horses: Systematics, Paleobiology, and Evolution of the Family Equidae*. Cambridge University Press, 1994.

Mair, V. H. The Horse in Late Prehistoric China: Wrestling Culture and Control from the "Barbarians." In *Prehistoric Steppe Adaptation and the Horse*, M. Levine, C. Renfrew, and K. Boyle, Eds., pp. 163-87. McDonald Institute for Archaeological Research, 2003.

Makarewicz, C. A., Winter-Schuh, C., Byerly, H. and Houle, J.-L. Isotopic Evidence for Ceremonial Provisioning of Late Bronze Age Khirigsuurs with Horses from Diverse Geographic Locales. *Quaternary International*, vol. 476, pp. 70-81, 2018. https://doi.org/10.1016/j.quaint.2018.02.030.

Mallen, L., Pearce, D., Arthur, C. and Mitchell, P. The Rock Arts of Metolong: Paintings, Archaeology and Cultural Resource Management in Western Lesotho. *Journal of African Archaeology*, vol. 20, no. 2, pp. 176-201, June 8, 2022.

Mallory-Greenough, L. The Horse Burials of Nubia. *Journal of the Society for the Study of Egyptian Antiquities*, vol. 32, pp. 105-10, 2005.

Maran, J. and de Moortel, A. V. A Horse-Bridle Piece with Carpatho-Danubian Connections from Late Helladic I Mitrou and the Emergence of a Warlike Elite in Greece during the Shaft Grave Period. *American Journal of Archaeology*, vol. 118, no. 4, p. 529, 2014.

Martin, L. and Russell, N. The Equid Remains from Neolithic Catalhoyuk, Central Anatolia: A Preliminary Report. In *Horses and Humans: The Evolution of Human-Equine Relationships*, S. Olsen, S. Grant, A. Choyke, and L. Bartosiewicz, Eds., pp. 115-26. British Archaeological Reports, 2006.

Martinić Beros, M. *Los Aónikenk: Historia y Cultura*. Universidad de Magallanes, 1995.

Marzahn, J. Equids in Mesopotamia—A Short Ride through Selected Textual Sources. In *Equids and Wheeled Vehicles in the Ancient World: Essays in Memory of Mary A. Littauer*, P. Raulwing, K. Linduff, and J. Crouwel, Eds., pp. 71-85. British Archaeological Reports, 2019.

Massilani, D., Skov, L., Hajdinjak, M., Gunchinsuren, B., Tseveendorj, D., Yi, S., Lee, J., et al. Denisovan Ancestry and Population History of Early East Asians. *Science*, vol. 370, no. 6516, pp. 579-83, October 30, 2020.

Maurer, G. and Greenberg, R. Cattle Drivers from the North? Animal Economy of a Diasporic Kura-Araxes Community at Tel Bet Yerah. *Levantina*, vol. 54, no. 3, pp. 309–30, September 2, 2022.

Mayor, A. *The Amazons*. Princeton University Press, 2014.

Mazzanti, D. and Quintana, C. Estrategias de Subsistencia de las Jefaturas Indígenas del Siglo XVIII: Zooarqueología de la Localidad Arqueológica Amalia (Tandilia Oriental). *Relaciones de la Sociedad Argentina de Antropología*, vol. 35, pp. 143–70, n.d.

McCormick, M., Büntgen, U., Cane, M. A., Cook, E. R., Harper, K., Huybers, P., Litt, T., et al. Climate Change during and after the Roman Empire: Reconstructing the Past from Scientific and Historical Evidence. *Journal of Interdisciplinary History*, vol. 43, no. 2, pp. 169–220, 2012.

McGovern, T. H., Smiarowski, K., Hambrecht, G., Brewington, S., Harrison, R., Hicks, M., Feeley, F., Prehal, B. and Woollett, J. Zooarchaeology of the Scandinavian Settlements in Iceland and Greenland. In *The Oxford Handbook of Zooarchaeology*, U. Albarella, H. Russ, K. Vickers, and S. Viner-Daniels, Eds. Oxford University Press, 2017.

McHorse, B. K., Biewener, A. A. and Pierce, S. E. Mechanics of Evolutionary Digit Reduction in Fossil Horses (Equidae). *Proceedings. Biological Sciences/The Royal Society*, vol. 284, no. 1861, August 30, 2017. https://doi.org/10.1098/rspb.2017.1174.

McHorse, B. K., Davis, E. B., Scott, E. and Jenkins, D. L. What Species of Horse Was Coeval with North America's Earliest Humans in the Paisley Caves? *Journal of Vertebrate Paleontology*, vol. 36, no. 6, p. e1214595, November 1, 2016.

Meyer, F. and Franke, P. R. Pferdetransport zur See im Altertum und Mittelalter. *Pferdeheilkunde*, vol. 20, no. 1, pp. 43–49, 2004.

Miller, B. K., Furholt, M., Bayarsaikhan, J., Tüvshinjargal, T., Brandtstätter, L., Wright, J., Ayush, T., et al. Proto-Urban Establishments in Inner Asia: Surveys of an Iron Age Walled Site in Eastern Mongolia. *Journal of Field Archaeology*, vol. 44, no. 4, pp. 267–86, May 19, 2019.

Miller, N. F., Spengler, R. N. and Frachetti, M. Millet Cultivation across Eurasia: Origins, Spread, and the Influence of Seasonal Climate. *Holocene*, vol. 26, no. 10, pp. 1566–75, 2016.

Minetti, A. E. Physiology: Efficiency of Equine Express Postal Systems. *Nature*, vol. 426, no. 6968, pp. 785–86, December 18, 2003.

Mitchell, M. D. Tracing Comanche History: Eighteenth Century Rock Art
    Depictions of Leather Armoured Horses from the Arkansas River Basin,
    South-Eastern Colorado, USA. *Antiquity*, vol. 78, no. 299, pp. 115–26,
    March 2004.

Mitchell, P. The Constraining Role of Disease on the Spread of Domestic
    Mammals in Sub-Saharan Africa: A Review. *Quaternary International*, vol.
    471, pp. 95–110, March 25, 2018.

Mitchell, P. *The Donkey in Human History: An Archaeological Perspective*.
    Oxford University Press, 2018.

Mitchell, P. *Horse Nations: The Worldwide Impact of the Horse on Indigenous
    Societies Post-1492*. Oxford University Press, 2015.

Mitchell, P. "A Horse Race Is the Same All the World Over": The Cultural
    Context of Horse Racing in Native North America. *International Journal of
    the History of Sport*, vol. 37, no. 3-4, pp. 337–56, March 3, 2020.

Mlinar, M. and Gerbec, T. *Keltskih Konj Topòt. Najdišče Bizjakova Hiša v
    Kobaridu/Hear the Horses of Celts. The Bizjakova Hiša Site in Kobarid
    Exhibition Catalogue*. Tolmin Museum, 2012.

Mokrynin, V. P. Археология и история древнего и средневекового
    Кыргызстана: избранное [Archaeology and History of Ancient and
    Medieval Kyrgyzstan: Selected Pieces]. Ilim, 2010.

Moorey, P. R. S. The Emergence of the Light, Horse-Drawn Chariot in the
    Near-East c. 2000–1500 B.C. *World Archaeology*, vol. 18, no. 2, pp. 196–215,
    October 1986.

Moorey, P. R. S. Pictorial Evidence for the History of Horse-Riding in Iraq
    before the Kassite Period. *Iraq*, vol. 32, no. 1, pp. 36–50, 1970.

Moreno, E. J. and Videla, B. A. Rastreando Ausencias: La Hipótesis del
    Abandono del Uso de los Recursos Marinos en el Momento Ecuestre en la
    Patagonia Continental. *Magallania*, vol. 36, no. 2, pp. 91–104, 2008.

Morgan, K., Funkquist, P. and Nyman, G. The Effect of Coat Clipping on
    Thermoregulation during Intense Exercise in Trotters. *Equine Veterinary
    Journal*, Supplement, no. 34, pp. 564–67, September 2002.

Morris, E. From Horse Power to Horsepower. *Access Magazine*, vol. 1, no. 30,
    2007.

Motuzaite Matuzeviciute, G., Preece, R. C., Wang, S., Colominas, L.,
    Ohnuma, K., Kume, S., Abdykanova, A. and Jones, M. K. Ecology and
    Subsistence at the Mesolithic and Bronze Age Site of Aigyrzhal-2, Naryn

Valley, Kyrgyzstan. *Quaternary International*, vol. 437, pp. 35–49, May 5, 2017.

Mühl, S. "Metal Makes the Wheel Go Round": The Development and Diffusion of Studded-Tread Wheels in the Ancient Near East and the Old World. In *Athyrmata: Critical Essays on the Archaeology of the Eastern Mediterranean in Honour of E. Susan Sherratt*, Y. Galanakis, T. Wilkinson, and J. Bennet, Eds., pp. 159–76. Archaeopress, 2014.

Musters, G. C. *At Home with the Patagonians: A Year's Wanderings over Untrodden Ground from the Straits of Magellan to the Rio Negro*. J. Murray, 1873.

Narasimhan, V. M., Patterson, N., Moorjani, P., Rohland, N., Bernardos, R., Mallick, S., Lazaridis, I., et al. The Formation of Human Populations in South and Central Asia. *Science*, vol. 365, no. 6457, September 6, 2019. https://doi.org/10.1126/science.aat7487.

Näser, C. and Mazzetti, G. Of Kings and Horses: Two New Horse Skeletons from the Royal Cemetery at El-Kurru, Sudan. *Archaeology International*, vol. 23, no. 1, pp. 122–37, December 30, 2020.

Navarro, T. Análisis Arqueofaunístico del Sitio El Panteon 1 (Las Ovejas, Neuquén). *La Zaranda de Ideas. Revista de Jóvenes Investigadores en Arqueología*, vol. 14, no. 1, pp. 41–54, 2016.

Nistelberger, H. M., Pálsdóttir, A. H., Star, B., Leifsson, R., Gondek, A. T., Orlando, L., Barrett, J. H., et al. Sexing Viking Age Horses from Burial and Non-Burial Sites in Iceland Using Ancient DNA. *Journal of Archaeological Science*, vol. 101, pp. 115–22, 2019.

Niven, L. From Carcass to Cave: Large Mammal Exploitation during the Aurignacian at Vogelherd, Germany. *Journal of Human Evolution*, vol. 53, no. 4, pp. 362–82, October 2007.

Noble, D. The Mesopotamian Onager as a Draught Animal. In *The Domestication and Exploitation of Plants and Animals*, P. Ucko and G. Dimbleby, Eds., pp. 485–88. Routledge, 2008.

Nordeide, S. W. The Oseberg Ship Burial in Norway: Introduction. *Acta Archaeologica*, vol. 82, no. 1, pp. 7–11, 2011.

O'Connor, S., McWilliam, A. and Brockwell, S. *Forts and Fortification in Wallacea: Archaeological and Ethnohistoric Investigations*. Australian National University Press, 2020.

Ogundiran, A. The Formation of an Oyo Imperial Colony during the Atlantic Age. In *Power and Landscape in Atlantic West Africa: Archeological*

Perspectives, J. Cameron Monroe and A. Ogundiran, Eds., pp. 222–52. Cambridge University Press, 2012.

Ogundiran, A. *The Yoruba: A New History*. Indiana University Press, 2020.

Olsen, S. L. Early Horse Domestication: Weighing the Evidence. In *Horses and Humans: The Evolution of Human-Equine Relationships*, S. Olsen, S. Grant, A. Choyke, and L. Bartosiewicz, Eds., pp. 81–114. British Archaeological Reports, 2006.

Olsen, S. L. The Exploitation of Horses at Botai, Kazakhstan. In *Prehistoric Steppe Adaptation and the Horse*, M. Levine, C. Renfrew, and K. Boyle, Eds., pp. 83–104. McDonald Institute for Archaeological Research, 2003.

Olsen, S. L. The Role of Humans in Horse Distribution through Time. In *Wild Equids: Ecology, Management, and Conservation*, J. I. Ransom and P. Kaczensky, Eds., pp. 105–20. Johns Hopkins University Press, 2016.

Olsen, S. L. Solutré: A Theoretical Approach to the Reconstruction of Upper Palaeolithic Hunting Strategies. *Journal of Human Evolution*, vol. 18, no. 4, pp. 295–327, June 1, 1989.

Olsen, S. L. and Culbertson, C. *A Gift from the Desert: The Art, History, and Culture of the Arabian Horse*. International Museum of the Horse, Kentucky Horse Park, 2010.

Orlando, L. Ancient Genomes Reveal Unexpected Horse Domestication and Management Dynamics. *BioEssays: News and Reviews in Molecular, Cellular and Developmental Biology*, vol. 42, no. 1, p. e1900164, January 2020.

Orlando, L., Ginolhac, A., Zhang, G., Froese, D., Albrechtsen, A., Stiller, M., Schubert, M., et al. Recalibrating Equus Evolution Using the Genome Sequence of an Early Middle Pleistocene Horse. *Nature*, vol. 499, no. 7456, pp. 74–78, July 4, 2013.

Orlando, L., Mashkour, M., Burke, A., Douady, C. J., Eisenmann, V. and Hänni, C. Geographic Distribution of an Extinct Equid (*Equus hydruntinus*: Mammalia, Equidae) Revealed by Morphological and Genetical Analyses of Fossils. *Molecular Ecology*, vol. 15, no. 8, pp. 2083–93, July 2006.

O'Shea, J. M. *Archaeology and Ethnohistory of the Omaha Indians: The Big Village Site*. University of Nebraska Press, 1992.

Ó Súilleabháin, S. Foundation Sacrifices. *Journal of the Royal Society of Antiquaries of Ireland*, vol. 75, no. 1, pp. 45–52, 1945.

Outram, A. K., Stear, N. A., Bendrey, R., Olsen, S., Kasparov, A., Zaibert, V., Thorpe, N., et al. The Earliest Horse Harnessing and Milking. *Science*, vol. 323, no. 5919, pp. 1332–35, March 6, 2009.

Outram, A. K., Stear, N. A., Kasparov, A., Usmanova, E., Varfolomeev, V. and Evershed, R. P. Horses for the Dead: Funerary Foodways in Bronze Age Kazakhstan. *Antiquity*, vol. 85, no. 327, pp. 116–28, March 2011.

Pedersen, A. Riding Gear from Late Viking-Age Denmark. *Journal of Danish Archaeology*, vol. 13, pp. 133–60, 1997.

Pederson, N., Hessl, A. E., Baatarbileg, N., Anchukaitis, K. J. and Di Cosmo, N. Pluvials, Droughts, the Mongol Empire, and Modern Mongolia. *Proceedings of the National Academy of Sciences of the United States of America*, vol. 111, no. 12, pp. 4375–79, March 25, 2014.

Perri, A. R., Feuerborn, T. R., Frantz, L. A. F., Larson, G., Malhi, R. S., Meltzer, D. J. and Witt, K. E. Dog Domestication and the Dual Dispersal of People and Dogs into the Americas. *Proceedings of the National Academy of Sciences of the United States of America*, vol. 118, no. 6, February 9, 2021. https://doi.org/10.1073/pnas.2010083118.

Perri, A., Widga, C., Lawler, D., Martin, T., Loebel, T., Farnsworth, K., Kohn, L. and Buenger, B. New Evidence of the Earliest Domestic Dogs in the Americas. *American Antiquity*, vol. 84, no. 1, pp. 68–87, January 2019.

Perşoiu, A., Onac, B. P., Wynn, J. G., Blaauw, M., Ionita, M. and Hansson, M. Holocene Winter Climate Variability in Central and Eastern Europe. *Scientific Reports*, vol. 7, no. 1, p. 1196, April 26, 2017.

Petrie, H. Satisfaction in a Horse: The Perception and Assimilation of an Exotic Animal into Maori Custom Law. In *Invasive and Introduced Plants and Animals*, I. Rotherham and R. Lambert, Eds., pp. 329–42. 2012.

Philipps, D. *Wild Horse Country: The History, Myth, and Future of the Mustang, America's Horse*. W. W. Norton, 2017.

Piggott, S. Chariots in the Caucasus and in China. *Antiquity*, vol. 48, no. 189, pp. 16–24, March 1974.

Pilø, L., Finstad, E. and Barrett, J. H. Crossing the Ice: An Iron Age to Medieval Mountain Pass at Lendbreen, Norway. *Antiquity*, vol. 94, no. 374, pp. 437–54, April 2020.

Podobed, V., Usachuk, A. and Tsimidanov, V. Cheek-Pieces of the Water Horses (One of the Mythologems in Cultures of Eurasia of the Bronze Age). In *Connections, Contacts and Interactions between Ancient Cultures of*

Northern Eurasia and Civilizations of the East during the Palaeometal Period (IV–I Mil. BC), A. V. Polyakov and E. S. Tkach, Eds., pp. 44–47. Institute of History and Material Culture (RAS), 2019.

Poliakov, A. V. and Svyatko, S. Modern Data on the Bronze Age Radiocarbon Chronology in the Minusinsk Basins. *Vestnik of Saint Petersburg University History*, vol. 66, no. 3, pp. 934–49, 2021.

Pope, M. and Parfitt, S. *The Horse Butchery Site: A High Resolution Record of Lower Palaeolithic Hominin Behaviour at Boxgrove, UK*. Spoilheap, 2020.

Price, N. The Vikings in Spain, North Africa and the Mediterranean. In *The Viking World*, S. Brink and N. Prince, Eds., pp. 486–93. Routledge, 2008.

Pruvost, M., Bellone, R., Benecke, N., Sandoval-Castellanos, E., Cieslak, M., Kuznetsova, T., Morales-Muñiz, A., et al. Genotypes of Predomestic Horses Match Phenotypes Painted in Paleolithic Works of Cave Art. *Proceedings of the National Academy of Sciences of the United States of America*, vol. 108, no. 46, pp. 18626–30, November 15, 2011.

Pryor, J. H. Transportation of Horses by Sea during the Era of the Crusades: Eighth Century to 1285 A. D. Part I, To c 1225. *The Mariner's Mirror*, vol. 68, no. 1, pp. 9–27, January 1, 1982.

Pustovalov, S. Курган «Тягунова Могила» и проблемы колесного транспорта ямно-катакомбной эпохи в Восточной Европе [The Tjagunova Mogila Burial Mound and the Problem of Wheeled Transport of the Pit Grave and Catacomb Cultures Epoch in Eastern Europe]. *Stratum Plus*, vol. 2, pp. 296–321, 2000.

Putnam, A. E., Putnam, D. E., Andreu-Hayles, L., Cook, E. R., Palmer, J. G., Clark, E. H., Wang, C., et al. Little Ice Age Wetting of Interior Asian Deserts and the Rise of the Mongol Empire. *Quaternary Science Reviews*, vol. 131, pp. 33–50, January 1, 2016.

Rannamäe, E., Andrianov, V., Järv, E., Semjonov, A., Haak, A. and Kreem, J. A Month in a Horse's Life: Healing Process of a Fractured Third Metatarsal Bone from Medieval Viljandi, Estonia. *International Journal of Paleopathology*, August 23, 2018. https://doi.org/10.1016/j.ijpp.2018.07.003.

Ransom, J. I. and Kaczensky, P. *Wild Equids: Ecology, Management, and Conservation*. Johns Hopkins University Press, 2016.

Rassamakin, Y. The Eneolithic of the Black Sea Steppe: Dynamics of Cultural and Economic Development 4500–2300 BC. In *Late Prehistoric Exploita-*

tion of the Eurasian Steppe, pp. 59–182. McDonald Institute for Archaeological Research, 1999.

Raulwing, P. *Horses, Chariots and Indo-Europeans*. Archaeolingua, 2000.

Rawson, J., Chugunov, K., Grebnev, Y. and Huan, L. Chariotry and Prone Burials: Reassessing Late Shang China's Relationship with Its Northern Neighbours. *Journal of World Prehistory*, 33, 135–68, July 2, 2020. https://link.springer.com/article/10.1007/s10963-020-09142-4.

Rawson, J., Huan, L. and Taylor, W. T. T., Seeking Horses: Allies, Clients and Exchanges in the Zhou Period (1045–221 BC). *Journal of World Prehistory*, December 24, 2021. https://doi.org/10.1007/s10963-021-09161-9.

Recht, L. *The Spirited Horse: Equid–Human Relations in the Bronze Age Near East*. Bloomsbury, 2022.

Reed, M. Horses in Pawnee History and Culture, Horses in the North American West. University of Colorado Museum of Natural History, September 1, 2021. https://www.colorado.edu/cumuseum/horses-north-american-west.

Reich, C. The Cemetery of Oberhof (Aukštkiemiai)—Horse Graves and Equestrian Equipment. *Archaeologia BALTICA*, vol. 11, pp. 206–16, 2009.

Reinhold, S., Gresky, J., Berezina, N., Kantorovich, A. R., Knipper, C., Maslov, V. E., Petrenko, et al. Contextualising Innovation. Cattle Owners and Wagon Drivers in the North Caucasus and Beyond. In *Appropriating Innovations: Entangled Knowledge in Eurasia 5000–1500 BCE*, J. Maran and P. Stockhammer, Eds., pp. 78–97. Oxbow, 2017.

Renton, K. E. A Social and Environmental History of the Horse in Spain and Spanish America, 1492–1600. PhD diss., University of California Los Angeles, 2018.

Richards, J. *The Secret War: A True History of Queensland's Native Police*. University of Queensland Press, 2008.

Rivallain, J. The Horse, the Status Mount of Africa. In *The Horse in the Art of the Near East*, D. Alexander, Ed., pp. 216–21. Vol. 1 of *Furusiyya: The Horse in the Art of the Near East*. King Abdulaziz Public Library, 1996.

Robin, C. Sabean and Himyarites Discover the Horse. In *The Horse in the Art of the Near East*, D. Alexander, Ed., pp. 60–72. Vol. 1 of *Furusiyya: The Horse in the Art of the Near East*. King Abdulaziz Public Library, 1996.

Roe, F. G. *The Indian and the Horse*. University of Oklahoma Press, 1968.

Rogers, J. D., Ulambayar, E. and Gallon, M. Urban Centres and the Emergence of Empires in Eastern Inner Asia. *Antiquity*, vol. 79, no. 306, pp. 801–18, December 2005.

Root, F. A. *The Overland Stage to California: Personal Reminiscences and Authentic History of the Great Overland Stage Line and Pony Express from the Missouri River to the Pacific Ocean*. F. A. Root and W. E. Connelley, 1901.

Rosenbom, S., Costa, V., Chen, S., Khalatbari, L., Yusefi, G. H., Abdukadir, A., Yangzom, C., et al. Reassessing the Evolutionary History of Ass-Like Equids: Insights from Patterns of Genetic Variation in Contemporary Extant Populations. *Molecular Phylogenetics and Evolution*, vol. 85, pp. 88–96, April 2015.

Rossel, S., Marshall, F., Peters, J., Pilgram, T., Adams, M. D. and O'Connor, D. Domestication of the Donkey: Timing, Processes, and Indicators. *Proceedings of the National Academy of Sciences of the United States of America*, vol. 105, no. 10, pp. 3715–20, March 11, 2008.

Rudenko, S. *Frozen Tombs of Siberia: The Pazyryk Burials of Iron-Age Horsemen*. University of California Press, 1970.

Sadykov, T., Caspari, G. and Blochin, J. Kurgan Tunnug 1—New Data on the Earliest Horizon of Scythian Material Culture. *Journal of Field Archaeology*, vol. 45, no. 8, pp. 556–70, November 16, 2020.

Sagona, A. *The Archaeology of the Caucasus: From Earliest Settlements to the Iron Age*. Cambridge University Press, 2018.

Sandall, R., dir. *Coniston Muster: Scenes from a Stockman's Life*. Film short. Australian Institute of Aboriginal Studies, 1972.

Sasaki, K. Adoption of the Practice of Horse-Riding in Kofun Period Japan: With Special Reference to the Case of the Central Highlands of Japan. *Japanese Journal of Archaeology*, vol. 6, pp. 23–53, 2018.

Sauer, J. J. *The Archaeology and Ethnohistory of Araucanian Resilience*. Springer International, 2014.

Sauvet, G. The Hierarchy of Animals in the Paleolithic Iconography. *Journal of Archaeological Science: Reports*, vol. 28, p. 102025, December 1, 2019.

Schaffer, I. *Land Musters, Stock Returns and Lists, Van Diemen's Land 1803–1822*. St. David's Park, 1991.

Schauensee, R. M. de. Horse Gear from Hasanlu. *Expedition*, vol. 31, no. 2–3, pp. 37–52, January 1, 1989.

Schmaus, T.M. Animals, Households, and Communities in Bronze and Iron Age Central Eurasia. In *Archaeologies of Animal Movement: Animals on the Move*, A.-K. Salmi and S. Niinimäki, Eds., pp. 85–93. Springer International, 2021.

Schrader, S.A., Smith, S.T., Olsen, S. and Buzon, M. Symbolic Equids and Kushite State Formation: A Horse Burial at Tombos. *Antiquity*, vol. 92, no. 362, pp. 383–97, April 2018.

Schroeder, B. The Alcova Redoubt: A Refuge Fortification in Central Wyoming. In *Archaeological Perspectives on Warfare on the Great Plains*, A. Clark and D. Bamforth, Eds., pp. 237–66. University Press of Colorado, 2018.

Schulman, A.R. Egyptian Representations of Horsemen and Riding in the New Kingdom. *Journal of Near Eastern Studies*, vol. 16, no. 4, pp. 263–71, October 1, 1957.

Scott, A., Reinhold, S., Hermes, T., Kalmykov, A.A., Belinskiy, A., Buzhilova, A., Berezina, N., et al. Emergence and Intensification of Dairying in the Caucasus and Eurasian Steppes. *Nature Ecology & Evolution*, vol. 6, no. 6, pp. 813–22, June 2022.

Seguin-Orlando, A., Donat, R., Der Sarkissian, C., Southon, J., Thèves, C., Manen, C., Tchérémissinoff, Y., et al. Heterogeneous Hunter-Gatherer and Steppe-Related Ancestries in Late Neolithic and Bell Beaker Genomes from Present-Day France. *Current Biology*, vol. 31, no. 5, pp. 1072–83.e10, March 8, 2021.

Seitsonen, O., Broderick, L., Houle, J.-L. and Bayarsaikhan, J. The Mystery of the Missing Caprines: Stone Circles at the Great Khirigsuur in the Khanuy Valley. *Studia Archaeologica*, no. 34, pp. 164–74, 2014.

Shackleton, S.E.H. *The Heart of the Antarctic: Being the Story of the British Antarctic Expedition 1907–1909.* J.B. Lippincott, 1914.

Shai, I., Greenfield, H.J., Brown, A., Albaz, S. and Maeir, A.M. The Importance of the Donkey as a Pack Animal in the Early Bronze Age Southern Levant: A View from "Tell Eṣ-Ṣāfī"/Gath. *Zeitschrift Des Deutschen Palästina-Vereins*, vol. 132, no. 1, pp. 1–25, 2016.

Shchetenko, A. Время появления домашней лошади (*Equus caballus*) в Средней Азии [The Time of the Appearance of the Domesticated Horse (*Equus caballus*) in Central Asia]. In Происхождение и распространение колес ничества сборник научных статей [Origin and Spreading of

Chariots—Collection of Scientific Articles], Vasilenko, Ed., pp. 219-33. Globus, 2008.

Shelach-Lavi, G., Jaffe, Y. and Bar-Oz, G. Cavalry and the Great Walls of China and Mongolia. *Proceedings of the National Academy of Sciences of the United States of America*, vol. 118, no. 16, e2024835118, 2021.

Shenk, P. To Valhalla by Horseback? Horse Burial in Scandinavia during the Viking Age. Master's thesis, University of Oslo, 2002.

Shim, H. The Postal Roads of the Great Khans in Central Asia under the Mongol-Yuan Empire. *Journal of Song-Yuan Studies*, vol. 44, no. 1, pp. 405-69, 2014.

Shishlina, N. I., Kovalev, D. S. and Ibragimova, E. R. Catacomb Culture Wagons of the Eurasian Steppes. *Antiquity*, vol. 88, no. 340, pp. 378-94, June 2014.

Singleton, J. Britain's Military Use of Horses, 1914-1918. *Past & Present*, no. 139, pp. 178-203, 1993.

Skvortsov, K. Burials of Riders and Horses Dated to the Roman Iron Age and Great Migration Period in Aleika-3 (Former Jaugehnen), Cemetery on the Sambian Peninsula. *Archaeologia Baltica*, vol. 11, pp. 130-48, 2009.

Smiarowski, K. Climate-Related Farm-to-Shieling Transition at E74 Qorlortorsuaq in Norse Greenland. In *Human Ecodynamics in the North Atlantic: A Collaborative Model of Humans and Nature through Space and Time*, R. Harrison and R. Maher, Eds., pp. 177-94. Lexington Books, 2014.

Spengler, R., Frachetti, M., Doumani, P., Rouse, L., Cerasetti, B., Bullion, E. and Mar'yashev, A. Early Agriculture and Crop Transmission among Bronze Age Mobile Pastoralists of Central Eurasia. *Proceedings. Biological Sciences/The Royal Society*, vol. 281, no. 1783, p. 20133382, May 22, 2014.

Spengler, R. N. III, Miller, A. V., Schmaus, T., Matuzevičiūtė, G. M., Miller, B. K., Wilkin, S., Taylor, W. T. T., et al. An Imagined Past? Nomadic Narratives in Central Asian Archaeology. *Current Anthropology*, vol. 62, no. 3, pp. 251-86, June 1, 2021.

Spyrou, M. A., Keller, M., Tukhbatova, R. I., Scheib, C. L., Nelson, E. A., Andrades Valtueña, A., Neumann, G. U., et al. Phylogeography of the Second Plague Pandemic Revealed through Analysis of Historical Yersinia Pestis Genomes. *Nature Communications*, vol. 10, no. 1, p. 4470, October 2, 2019.

Spyrou, M. A., Tukhbatova, R. I., Wang, C.-C., Valtueña, A. A., Lankapalli, A. K., Kondrashin, V. V., Tsybin, V. A., et al. Analysis of 3800-Year-Old Yersinia Pestis Genomes Suggests Bronze Age Origin for Bubonic Plague. *Nature Communications*, vol. 9, no. 1, p. 2234, June 8, 2018.

Stark, S., Rubinson, K. S. and Samashev, Z. *Nomads and Networks: The Ancient Art and Culture of Kazakhstan*. Princeton University Press, 2012.

Steele, J. and Politis, G. AMS 14C Dating of Early Human Occupation of Southern South America. *Journal of Archaeological Science*, vol. 36, no. 2, pp. 419–29, February 1, 2009.

Stepanova, E., Chugunov, K. Horse Equipment of the Late Bronze Age in Early China: Revolution and Evolution. In *Connections, Contacts and Interactions between Ancient Cultures of Northern Eurasia and Civilizations of the East during the Palaeometal Period (IV–I Mil. BC)*, A. V. Polyakov and E. S. Tkach, Eds., pp. 95–97. Institute of History and Material Culture (RAS), 2019.

Stepanova, E. V. Saddles of the Hun-Sarmatian Period. In *Masters of the Steppe: The Impact of the Scythians and Later Nomad Societies of Eurasia*, S. Pankova and S. Simpson, Eds., pp. 561–88. Archaeopress, 2021.

Steppan, K. The Neolithic Human Impact and Wild Horses in Germany and Switzerland: Horse Size Variability and the Chrono-Ecological Context. In *Horses and Humans: The Evolution of Human-Equine Relationships*, S. Olsen, S. Grant, A. Choyke, and L. Bartosiewicz, Eds., pp. 209–20. British Archaeological Reports, 2006.

Strassnig, C. Rediscovering the Camino Real of Panama: Archaeology and Heritage Tourism Potentials. *Journal of Latin American Geography*, vol. 9, no. 2, pp. 159–68, 2010.

Straus, L. G. Upper Paleolithic Hunting Tactics and Weapons in Western Europe, *Archaeological Papers of the American Anthropological Association*, vol. 4, no. 1, pp. 83–93, January 1, 1993.

Street, J. M. Feral Animals in Hispaniola. *Geographical Review*, vol. 52, no. 3, pp. 400–6, 1962.

Strömberg, C. A. E. Evolution of Grasses and Grassland Ecosystems. *Annual Review of Earth and Planetary Sciences*, vol. 39, pp. 517–44, 2011. https://doi.org/10.1038/s41598-022-06659-w.

Struck, J., Bliedtner, M., Strobel, P., Taylor, W. T. T., Biskop, S., Plessen, B., Klaes, B., et al. Climate Change and Equestrian Empires in the Eastern

Steppes: New Insights from a High-Resolution Lake Core in Central Mongolia. *Scientific Reports*, vol. 12, no. 2829, 2022.

Su, Y., Fang, X. and Yin, J. Impact of Climate Change on Fluctuations of Grain Harvests in China from the Western Han Dynasty to the Five Dynasties (206 BC–960 AD). *Science China Earth Sciences*, vol. 57, no. 7, pp. 1701–12, July 1, 2014.

Sundstrom, L. Coup Counts and Corn Caches: Contact-Era Plains Indian Accounts of Warfare. In *Archaeological Perspectives on Warfare on the Great Plains*, A. Clark and D. Bamforth, Eds., pp. 120–42. University of Colorado Press, 2018.

Sussman, R. W., Rasmussen, D. T. and Raven, P. H. "Rethinking Primate Origins Again." *American Journal of Primatology*, vol. 75, no. 2, pp. 95–106, 2013.

Svyatko, S. V., Mallory, J. P., Murphy, E. M., Polyakov, A. V., Reimer, P. J. and Schulting, R. J. New Radiocarbon Dates and a Review of the Chronology of Prehistoric Populations from the Minusinsk Basin, Southern Siberia, Russia. *Radiocarbon*, vol. 51, no. 1, pp. 243–73, 2009.

Swart, S. *Riding High: Horses, Humans and History in South Africa*. New York University Press, 2010.

Tabaldyev, K. Monuments of the Bronze Age of Kyrgyzstan. *Himalayan and Central Asian Studies*, vol. 15, no.1, pp. 3–12, 2012.

Takács, I. Evidence of Horse Use and Harnessing on Horse Skeletons from the Migration Period and the Time of the Hungarian Conquest. *Archaeozoologia*, vol. 7, no. 2, pp. 43–53, 1995.

Tao, T. and Wertmann, P. The Coffin Paintings of the Tubo Period from the Northern Tibetan Plateau. *Bridging Eurasia*, vol. 1, pp. 187–213, 2010.

Tarasov, P. E., Jolly, D. and Kaplan, J. O. A Continuous Late Glacial and Holocene Record of Vegetation Changes in Kazakhstan. *Palaeogeography, Palaeoclimatology, Palaeoecology*, vol. 136, no. 1, pp. 281–92, December 15, 1997.

Taylor, W. Horse Demography and Use in Bronze Age Mongolia. *Quaternary International*, vol. 436, pp. 270–82, April 29, 2017.

Taylor, W. How Dan the Zebra Stopped an Ill-Fated Government Breeding Program in Its Tracks. *Smithsonian Magazine Online*, December 4, 2019. https://www.smithsonianmag.com/smithsonian-institution/how-dan -zebra-stopped-ill-fated-governent-breeding-program-tracks-180973542/.

Taylor, W. Pandemics and the Post: Mongolia's Pony Express. *Diplomat*, October 16, 2020. https://thediplomat.com/2020/10/pandemics-and-the-post-mongolias-pony-express/.

Taylor, W. and Barrón-Ortiz, C. I. Rethinking the Evidence for Early Horse Domestication at Botai. *Scientific Reports*, vol. 11, no. 1, p. 7440, April 2, 2021.

Taylor, W., Bayarsaikhan, J. and Tuvshinjargal, T. Equine Cranial Morphology and the Identification of Riding and Chariotry in Late Bronze Age Mongolia. *Antiquity*, vol. 89, no. 346, pp. 854–71, August 2015.

Taylor, W., Bayarsaikhan, J., Tuvshinjargal, T., Bender, S., Tromp, M., Clark, J., Lowry, K. B., et al. Origins of Equine Dentistry. *Proceedings of the National Academy of Sciences of the United States of America*, vol. 115, no. 29, pp. e6707-15, 2018.

Taylor, W., Belardi, J. B., Barbarena, R., Brenner Coltrain, J., Carballo Marina, F., Borrero, L. A., Conver, J. L., et al. Interdisciplinary evidence for early domestic horse exploitation in southern Patagonia. *Science Advances*, vol. 9, no. 49. https://doi.org/10.1126/sciadv.adk5201.

Taylor, W., Cao, J., Fan, W., Ma, X., Hou, Y., Wang, J., Li, Y., et al. Understanding Early Horse Transport in Eastern Eurasia through Analysis of Equine Dentition. *Antiquity*, vol. 95, no. 384, pp. 1478–94, 2021.

Taylor, W., Clark, J., Bayarsaikhan, J., Tuvshinjargal, T., Jobe, J. T., Fitzhugh, W., Kortum, R., et al. Early Pastoral Economies and Herding Transitions in Eastern Eurasia. *Scientific Reports*, vol. 10, no. 1, p. 1001, January 22, 2020.

Taylor, W., Clark, J. K., Reichhardt, B., Hodgins, G. W. L., Bayarsaikhan, J., Batchuluun, O., Whitworth, J., et al. Investigating Reindeer Pastoralism and Exploitation of High Mountain Zones in Northern Mongolia through Ice Patch Archaeology. *PloS One*, vol. 14, no. 11, p. e0224741, November 20, 2019.

Taylor, W., Fantoni, M., Marchina, C., Lepetz, S., Bayarsaikhan, J., Houle, J.-L., Pham, V., et al. Horse Sacrifice and Butchery in Bronze Age Mongolia. *Journal of Archaeological Science: Reports*, vol. 31, p. 102313, June 1, 2020.

Taylor, W., Hart, I., Jones, E. L., Brenner-Coltrain, J., Jobe, J. T., Britt, B. B., Gregory McDonald, H., et al. Interdisciplinary Analysis of the Lehi Horse: Implications for Early Historic Horse Cultures of the North American West. *American Antiquity*, vol. 86, no. 3, pp. 465–85, 2021.

Taylor, W., Jargalan, B., Lowry, K.B., Clark, J., Tuvshinjargal, T. and Bayar-saikhan, J. A Bayesian Chronology for Early Domestic Horse Use in the Eastern Steppe. *Journal of Archaeological Science*, vol. 81, Supplement C, pp. 49–58, May 1, 2017.

Taylor, W., Librado, P., American Horse, C.J., Shield Chief Gover, C., Arterberry, J., Afraid of Bear-Cook, A.L., Left Heron, H., et al. Early Dispersal of Domestic Horses into the Great Plains and Northern Rockies. *Science*, vol. 379, no. 6639, pp. 1316–23, March 31, 2023.

Taylor, W., Pruvost, M., Posth, C., Rendu, W., Krajcarz, M.T., Abdykanova, A., Brancaleoni, G., et al. Evidence for Early Dispersal of Domestic Sheep into Central Asia. *Nature Human Behaviour*, vol. 5, pp. 1169–79, 2021. https://doi.org/10.1038/s41562-021-01083-y.

Taylor, W. and Tuvshinjargal, T. Horseback Riding, Asymmetry, and Changes to the Equine Skull: Evidence for Mounted Riding in Mongolia's Late Bronze Age. In *Care or Neglect? Evidence of Animal Disease in Archaeology*, L. Bartosiewicz and E. Gal, Eds., p. 134–54. Oxbow, 2018.

Taylor, W., Tuvshinjargal, T. and Bayarsaikhan, J. Reconstructing Equine Bridles in the Mongolian Bronze Age. *Journal of Ethnobiology*, vol. 36, no. 3, pp. 554–70, October 1, 2016.

Todd, E.T., Tonasso-Calvière, L., Chauvey, L., Schiavinato, S., Fages, A., Seguin-Orlando, A., Clavel, P., et al. The Genomic History and Global Expansion of Domestic Donkeys. *Science*, vol. 377, no. 6611, pp. 1172–80, September 9, 2022.

Tozaki, T., Kikuchi, M., Kakoi, H., Hirota, K., Nagata, S., Yamashita, D., Ohnuma, T., et al. Genetic Diversity and Relationships among Native Japanese Horse Breeds, the Japanese Thoroughbred and Horses outside of Japan Using Genome-Wide SNP Data. *Animal Genetics*, vol. 50, no. 5, pp. 449–59, October 2019.

Trautmann, M., Frînculeasa, A., Preda-Bălănică, B., Petruneac, M., Focşăneanu, M., Alexandrov, S., Atanassova, N., et al. First Bioanthropo-logical Evidence for Yamnaya Horsemanship. *Science Advances*, vol. 9, no. 9, p. eade2451, March 3, 2023.

Troncoso, A., Pascual, D. and Moya, F. Making Rock Art under the Spanish Empire: A Comparison of Hunter Gatherer and Agrarian Contact Rock Art in North-Central Chile, *Australian Archaeology*, vol. 84, no. 3, pp. 263–80, September 2, 2018.

Turbat, T., Batsukh, D. and Bayarkhuu, N. Хүннүгийн археологийн тамгууд Люаньди овгийн тамга болох нь [Xiongnu Archaeological Tamgas as Luanti Clan Property Signs]. *Studia Archaeologica*, vol. 32, pp.136–61, 2012.

Turfan City Bureau of Cultural Relics, Xinjiang Institute of Cultural Relics and Archaeology, Academy of Turfanology and Turfan Museum. 新疆洋海墓地 [Report of Archaeological Excavations at Yanghai Cemetery]. 文物出版社 [Cultural Relics Press], 2019.

Tuvshinjargal, T. and Taylor, W. "Гэрийн Эртний Түүхэн" Хөгжлийг Археологийн Хэрэглэгдэхүүнээр Тодруулах Нь [Clarifying the Early Historical Development of the Ger in the Archaeological Record]. In Монгол Гэрийн Өв Соёл [The Cultural Heritage of the Mongolian Ger]. D. Sukhbaatar, Ed., pp. 5–13. National Museum of Mongolia, 2019.

Uetsuki, M., Gakuhari, T., Isahaya, N., Maruyama, M. and Aoyagai, T. Horse Feeding Strategy in Ancient Japan and Korea: Stable Carbon Isotope Analysis of the Tooth Enamel. 대한체질인류학회 학술대회 연제 초록 [Korean Society for Physical Anthropology], vol. 63, pp. 5–6, August 2020.

Uetsuki, M., Nishinakagawa, H. and Yamaji, N. The Use of Horses in Classical Period Japan Inferred from Pathology and Limb Bone Proportion. *Asian Journal of Paleopathology*, vol. 4, pp. 13–28, 2022. https://doi.org/10.32247/ajp2022.4.4.

Van Buren, E. D. *Clay Figurines of Babylonia and Assyria*. AMS Press, 1980.

Vander Velden, F. A Tapuya "Equestrian Nation"? Horses and Native Peoples in the Backlands of Colonial Brazil. In *The Materiality of the Horse*, M. Bibby and B. Scott, Eds., pp. 71–106. Trivent, 2023.

Vasil'ev, S. A. Faunal Exploitation, Subsistence Practices and Pleistocene Extinctions in Paleolithic Siberia. *Deinsea*, vol. 9, no. 1, pp. 513–56, January 1, 2003.

Vasiliev, S. K. Large Mammal Fauna from the Pleistocene Deposits of Chagyrskaya Cave Northwestern Altai (Based on 2007–2011 Excavations). *Archaeology, Ethnology and Anthropology of Eurasia*, vol. 41, no. 1, pp. 28–44, March 1, 2013.

Vershinina, A. O., Heintzman, P. D., Froese, D. G., Zazula, G., Cassatt-Johnstone, M., Dalén, L., Der Sarkissian, C., et al. Ancient Horse Genomes Reveal the Timing and Extent of Dispersals across the Bering Land Bridge. *Molecular Ecology*, vol. 30, no. 23, pp. 6144–61, December 2021.

Vigne, J.-D. Early Domestication and Farming: What Should We Know or Do for a Better Understanding? *Anthropozoologica*, vol. 50, no. 2, pp. 123–51, 2015.

Vila, E. Data on Equids from Late Fourth and Third Millennium Sites in Northern Syria. In *Equids in Time and Space: Papers in Honour of Vera Eisenmann*, M. Mashkour, Ed., pp. 101–23. Oxbow, 2006.

Villavicencio, N. A., Lindsey, E. L., Martin, F. M., Borrero, L. A., Moreno, P. I., Marshall, C. R. and Barnosky, A. D. Combination of Humans, Climate, and Vegetation Change Triggered Late Quaternary Megafauna Extinction in the Última Esperanza Region, Southern Patagonia, Chile. *Ecography*, vol. 39, no. 2, pp. 125–40, February 2016.

Vitt, V. O. The Horses of the Kurgans of Pazyryk. *Journal of Soviet Archaeology*, vol. 16, pp. 163–206, 1952.

Wagner, M., Bo, W., Tarasov, P., Westh-Hansen, S. M., Völling, E. and Heller, J. The Ornamental Trousers from Sampula (Xinjiang, China): Their Origins and Biography. *Antiquity*, vol. 83, no. 322, pp. 1065–75, December 2009.

Wagner, M., Wu, X., Tarasov, P., Aisha, A., Ramsey, C. B., Schultz, M., Schmidt-Schultz, T., et al. Radiocarbon-Dated Archaeological Record of Early First Millennium BC Mounted Pastoralists in the Kunlun Mountains, China. *Proceedings of the National Academy of Sciences*, vol. 108, no. 38, pp. 15733–38, 2011.

Wallace, B. L'Anse Aux Meadows, Leif Eriksson's Home in Vinland. *Journal of the North Atlantic*, vol. 2, no. sp2, pp. 114–25, October 2009.

Wallace, E. and Adamson Hoebel, E. *The Comanches: Lords of the South Plains*. University of Oklahoma Press, 2013.

Waters, M. R., Stafford, T. W., Jr, Kooyman, B. and Hills, L. V. Late Pleistocene Horse and Camel Hunting at the Southern Margin of the Ice-Free Corridor: Reassessing the Age of Wally's Beach, Canada. *Proceedings of the National Academy of Sciences of the United States of America*, vol. 112, no. 14, pp. 4263–67, April 7, 2015.

Wayland, V., Wayland, H. and Ferg, A. *Playing Cards of the Apaches: A Study in Cultural Adaptation*. Screenfold, 2006.

Weatherford, J. *Genghis Khan and the Quest for God: How the World's Greatest Conqueror Gave Us Religious Freedom*. Penguin, 2017.

Webb, S. D. and Hemmings, C. A. Last Horses and First Humans in North America. In *Horses and Humans: The Evolution of Human-Equine Relation-*

ships, S. Olsen, S. Grant, A. Choyke, and L. Bartosiewicz, Eds., pp. 11–24. British Archaeological Reports, 2006.

Weber, J. A. Elite Equids: Redefining Equid Burials of the Mid- to Late 3rd Millennium BC from Umm el-Marra, Syria. *MOM Éditions*, vol. 49, no. 1, pp. 499–519, 2008.

Weber, J. A., Porter, A. and Schwartz, G. Restoring Order: Death, Display and Authority. In *Sacred Killing: The Archaeology of Sacrifice in the Ancient Near East*, A. Porter and G. Schwartz, Eds., pp. 159–90. Eisenbraun, 2012.

Wedel, W. R. Coronado, Quivira, and Kansas: An Archeologist's View. *Great Plains Quarterly*, vol. 10, no. 3, pp. 139–51, 1990.

Wertmann, P., Chen, X., Li, X., Xu, D., Tarasov, P. E. and Wagner, M. New Evidence for Ball Games in Eurasia from ca. 3000-Year-Old Yanghai Tombs in the Turfan Depression of Northwest China. *Journal of Archaeological Science: Reports*, vol. 34, p. 102576, December 1, 2020.

Wertmann, P., Xu, D., Elkina, I., Vogel, R., Yibulayinmu, M. 'eryamu, Tarasov, P. E., La Rocca, D. J. and Wagner, M. No Borders for Innovations: A ca. 2700-Year-Old Assyrian-Style Leather Scale Armour in Northwest China. *Quaternary International*, vol. 623, pp. 110–26, 2021. https://doi .org/10.1016/j.quaint.2021.11.014.

Wertmann, P., Yibulayinmu, M., Wagner, M., Taylor, C., Müller, S., Xu, D., Elkina, I., Leipe, C., Deng, Y. and Tarasov, P. E. The Earliest Directly Dated Saddle for Horse-Riding from a Mid-1st Millennium BCE Female Burial in Northwest China. *Archaeological Research in Asia*, vol. 35, p. 100451, September 1, 2023.

Wheat, J. B., Malde, H. E. and Leopold, E. B. The Olsen-Chubbuck Site: A Paleo-Indian Bison Kill. *Memoirs of the Society for American Archaeology*, no. 26, pp. i–180, 1972.

Wilkin, S., Ventresca Miller, A., Fernandes, R., Spengler, R., Taylor, W. T.-T., Brown, D. R., Reich, D., et al. Dairying Enabled Early Bronze Age Yamnaya Steppe Expansions. *Nature*, vol. 598, no. 7882, pp. 629–33, October 2021.

Wilkin, S., Ventresca Miller, A., Miller, B. K., Spengler, R. N., III, Taylor, W. T. T., Fernandes, R., Hagan, R. W., et al. Economic Diversification Supported the Growth of Mongolia's Nomadic Empires. *Scientific Reports*, vol. 10, no. 1, p. 3916, March 3, 2020.

Wilkin, S., Ventresca Miller, A., Taylor, W. T. T., Miller, B. K., Hagan, R. W., Bleasdale, M., Scott, A., et al. Dairy Pastoralism Sustained Eastern Eurasian Steppe Populations for 5,000 Years. *Nature Ecology & Evolution*, vol. 4, no. 3, pp. 346–55, March 2020.

Willcox, G. and Stordeur, D. Large-Scale Cereal Processing before Domestication during the Tenth Millennium Cal BC in Northern Syria. *Antiquity*, vol. 86, no. 331, 99–114, 2012.

Williams, G. *Weapons of the Viking Warrior*. Bloomsbury, 2019.

Wilson, G. L. *The Horse and the Dog in Hidatsa Culture*. American Museum Press, 1924.

Wit, P. and Bouman, I. *The Tale of the Przewalski's Horse: Coming Home to Mongolia*. KNNV, 2006.

Wood, A. R., Bebej, R. M., Manz, C. L., Begun, D. L. and Gingerich, P. D., Postcranial Functional Morphology of Hyracotherium (Equidae, Perissodactyla) and Locomotion in the Earliest Horses. *Journal of Mammalian Evolution*, vol. 18, no. 1, pp. 1–32, March 2011.

Wright, J., Honeychurch, W. and Amartuvshin, C. The Xiongnu Settlements of Egiin Gol, Mongolia. *Antiquity*, vol. 83, no. 320, pp. 372–87, June 2009.

Wu, X. *Chariots in Early China: Origins, Cultural Interaction, and Identity*. British Archaeological Reports, 2013.

Wutke, S., Andersson, L., Benecke, N., Sandoval-Castellanos, E., Gonzalez, J., Hallsson, J. H., Lõugas, L., et al. The Origin of Ambling Horses. *Current Biology*, vol. 26, no. 15, pp. R697–99, August 8, 2016.

Xenophon. *The Art of Horsemanship*. J. M. Dent, 1894.

Yang, L., Kong, X., Yang, S., Dong, X., Yang, J., Gou, X. and Zhang, H. Haplotype Diversity in Mitochondrial DNA Reveals the Multiple Origins of Tibetan Horse. *PloS One*, vol. 13, no. 7, p. e0201564, July 27, 2018.

Yuan, J. and Flad, R. Research on Early Horse Domestication in China. In *Equids in Time and Space: Papers in Honour of Vera Eisenmann*, M. Mashkour, Ed., pp. 124–32. Oxbow, 2006.

Zahir, M. Gandhara Grave Culture: New Perspectives on Protohistoric Cemeteries in Northern and North-Western Pakistan. In *A Companion to South Asia in the Past*, G. R. Schug and S. R. Walimbe, Eds., pp. 274–93. Wiley, 2016.

Zarins, J. and Hauser, R. *The Domestication of Equidae in Third-Millennium BCE Mesopotamia*. CDL, 2014.

Zazzo, A., Lepetz, S., Magail, J. and Gantulga, J.-O. High-Precision Dating of Ceremonial Activity around a Large Ritual Complex in Late Bronze Age Mongolia. *Antiquity*, vol. 93, no. 367, pp. 80–98, February 2019.

Zeder, M. A. The Equid Remains from Tal-E Malyan, Southern Iran. In *Equids in the Ancient World*, R. H. Meadow and H. P. Uerpmann, Eds., pp. 366–412. Ludwig Reichert Verlag, 1986.

Zeder, M. A. Out of the Fertile Crescent: The Dispersal of Domestic Livestock through Europe and Africa. In *Human Dispersal and Species Movement: From Prehistory to the Present*, N. Boivin, R. Crassard, and M. Petraglia, Eds., pp. 261–303. Cambridge University Press, 2017.

Zeder, M. A. and Hesse, B. The Initial Domestication of Goats (Capra hircus) in the Zagros Mountains 10,000 Years Ago. *Science*, vol. 287, no. 5461, pp. 2254–57, March 24, 2000.

Zhang, C., Wang, Y., Zhang, J., Taylor, W. T. T., Sun, F., Huang, Z., Qiu, R., et al. Elite Chariots and Early Horse Transport at the Bronze Age Burial Site of Shijia. *Antiquity*, vol. 97, no. 393, pp. 636–53, 2023.

Zhang, D. and Feng, Z. Holocene Climate Variations in the Altai Mountains and the Surrounding Areas: A Synthesis of Pollen Records. *Earth-Science Reviews*, vol. 185, pp. 847–69, October 1, 2018.

Zhang, Z. W., Wangdue, S., Lu, H. L. and Nyima, S. C. Identification and Interpretation of Faunal Remains from a Prehistoric Cist Burial in Amdo County North Tibet. *Journal of Tibetology*, vol. 12, pp. 1–18, 2015.

INDEX

Abashevo culture (Romania), chariots of, 89

aboriginal people of Australia, 207; in European cavalry units, 210; genocide of, 210; relationship with domestic horses, 210–12; trade in horse tack, 210–11; of 21st century, 222

adaptive radiation, following K-Pg Extinction event, 5–6

Afanasievo culture: dairy products of, 62; herders, 63*fig.*, 64; livestock of, 61; in Mongolia, 104. *See also* Yamnaya culture

Africa: ancestral zebras of, 17; colonial exploitation of, 79; colonial society of, 206; equine diseases of, 165; grassland cultures, 165; horse-mediated linkages across, 100–101*map*; Neolithic livestock in, 165; primate evolution in, 7; rainforests, 164–65; wild asses of, 15

Africa, Horn of: introduction of horses to, 162

agriculture: Australian, 207–8; donkeys' use in, 66; horses' role in, 34, 178, 215*fig.*; lower temperatures affecting, 153–54; transport of crops, 98

airag (fermented milk), 34, 114

Akkadian language, chariot-related words in, 90

Alakul culture, Trans-Ural: chariot burials in, 88

Alberta (Canada), caballine horse remains of, 24–25

Alexander the Great, horse supply of, 136

Altai Mountains: Afanasievo herders of, 64; archaeological record of, 106–7; chariot burials in, 88; domestic horses of, 107, 133–34; herding cultures of, 106; horse hunting in, 19; livestock of, 61, 105; petroglyphs of, 86–87; population flow into Kazakhstan, 126. *See also* Pazyryk culture

Amazons, legends of, 131

ambling gait (*tölt*), Icelandic, 176

Barker, Collet, 209
Barrón-Ortiz, Christina, 52
Basotho people, Bantu-speaking: ritual use of horses, 206
battle cars, equid, 71*fig.*, 74
Bayarsaikhan, J., 151–52
Bay of Fundy, introduction of horses to, 187
Bedouins, horses of, 161
Biluut (Mongolia), livestock of, 105
bison, wild: drawing ox carts, 80. *See also* buffalo hunting
bits (bridles), 43–44; Arabian, 163; of Arzhan site, 124; curb, 150; damage to horses' teeth, 41–42; jointed snaffled, 123–24, 125*fig.*; metal, 122–23; Near Eastern, 91; oldest, 72; of Sintashta culture, 122. *See also* bridles; horse control
Blackfoot Confederacy (Great Plains), southward migration of, 191
Black Sea steppes: domestication of horses in, 76–77; horse transport in, 81; wild caballine taxon of, 75; Yamnaya ancestry in, 61
Blacks Fork, horse burial of, 186
Bluefish Caves (Yukon), archaeological assemblage of, 24
bola, hunting with, 199
Borobodur (Java), horse relief carvings of, 206
Botai (Kazakhstan): bit-wear damage at, 49; bone/antler projectiles, 48*fig.*; DNA from, 45, 51; domestication hypothesis of, 46, 48–53, 54; *Equus przewalskii* of, 51; horse remains at, 45–46, 48;

human DNA from, 51; human/horse burial at, 45; hunter-gatherer structures at, 53; hunting of horses at, 48, 53; pottery, 49–50; skeletal measurements at, 49; tooth damage at, 52
Boxgrove (United Kingdom), horse-hunting fossils of, 18–19
Brazil, introduction of horses to, 197
breeding, horse: human control over, 34, 36, 85, 188; Indigenous American/colonial, 192; for riding, 123–24; spinal column in, 81; of Viking horses, 176
bridles: cheekpieces, 44, 89, 94; effect on horses' skeletons, 42; identification, 44–45; Indigenous American, 188–89; of Karasuk culture, 106; new technologies, 124; Pazyryk, 133; Persian, 135; of Petrovka culture, 87–88; Shang dynasty, 117–18; in Sintashta culture, 83–84; south of the Caucasus, 90–91; in the steppes, 90, 123; trans-Saharan, 163; from Troy, 94; from Turkmenistan, 96. *See also* bits; horse control
British Isles: Anglo-Saxon invasion of, 174; chariots in, 89; transport of horses into, 171, 177; Vikings in, 176, 177
Bronze Age: chariot carving of, 113; horse herders of, 114–15; veterinary care in, 114–15
Brown, Dorcas, 42; on bridles, 45
Bucephalus (horse of Alexander the Great), 136
Buenos Aires, settlement of, 197

buffalo hunting, on horseback, 190–91

burials, horse: Baltic region, 174; of Blacks Fork, 186; at Botai (Kazakhstan), 45; demographics of, 41; of Fedorovo culture, 88, 105; of Gandhara Grave complex, 96; in Japan, 173; of Karasuk culture, 89; of *khirigsuurs*, 109; Late Helladic III period, 93; of Liushui, 112; of Minusinsk Basin, 105; mounds, 109–10, 115; of Napatan kingdoms, 160; Norse Icelandic, 177–78; of Pazyryk culture, 133–34; radiocarbon study of, 110; in Scandinavia, 174; of Sintashta culture, 83–84, 83*fig.*, 97; of Slovenia, 129; of Tibet, 146–47; Viking, 176, 177; of Xi'an, *plate 8*. *See also* sacrifices, horse; skeletons, equine

Butaxiongqu (Tibet), horse burial of, 147

buzkashi (Afghanistan), 222

caliphates, Arabian: cavalry of, 179; horse genome of, 162; horses of, 161–62; Rashidun, 162; technological innovations in, 161; Ummayad, 162

camelids, of Andean Empire, 195

camels: in desert warfare, 160; support for horse transport, 161, 214

Cameroon, introduction of horses to, 162

Camino de Cruces (Panama), 182

Camino Real, trans-isthmus, 182

Canary Islands, Castilian possession of, 180

Cape Fear (North Carolina), Ayllón expedition to, 183

Caribbean, introduction of horses to, 181–82, 183

Carthage, ancient: cavalry demands of, 146

carts: rein-and-ring system, 81, 83; safety of, 79; two-wheeled, 81–82, 82*fig.*; of Uttar Pradesh, 95; zebra, 80. *See also* wagons; wheels

carts, cattle: adaptation to donkeys, 78; construction of, 60–61; Maikop, 62; of Yamnaya culture, 62

Casa dei Casti Amanti (Pompei), horse remains from, 136, *plate 10*

Caspian region, stepped-linked culture of, 125–26

Catacomb culture, two-wheeled vehicles of, 81–82

cats, domestic versus feral, 30

cattle: carts, 60–62, 78; from Kura-Araxes culture, 60; nose rings for, 45, 60, *plate 5*; skeletal changes, 59

Caucus Mountains: cattle transport in, 60; domestic horses in, 90–91; stepped-linked culture of, 125–26

cavalry, 122, 124–26, *plate 7*; of Achaemenid Persia, 135–36; Assyrian, 125, 125*fig.*, 135; cultural interactions accompanying, 142–43; in defeat of Aztec Empire, 183; displacement of chariotry, 126; in emergence of steppe empire, 139–40; foot supports for,

cavalry *(continued)*
150; impact in Asia, 135–39; of
Mali, 164; North African, 160;
power associated with, 148; riders'
stability, 150, 152; riding in pairs,
124–25, 125*fig.*; Roman, 146, 150;
Saka, 135; Scythian, 135; wealth
from, 148; westward spread of,
128; women in, 130; during World
War I, 217; Xenophon on, 136. *See
also* riding; warfare
cecum, horses': micobiome of, 10–11
chamfron (horse's battle helmet),
126
chariots, 77, 81–98; in ancient
Mesopotamian languages, 90;
archaeozoology of, 87–88; archery
from, 86, 91; ascent to power
through, 92; in burials, 83–84,
83*fig.*, 88, 89, 97, 121; construction
materials, 86; on deer stone
monuments, 112; of DSK culture,
112–13; as elite conveyances, 86,
87, 92, 97, 117, 119–20, 121; in
Europe, 89, 93–94; in Greco-
Roman antiquity, 136–37; at
Hasanlu, 125–16; for herding, 86,
87, 130; Hittite, 91; impact on
authority systems, 93, 97–98;
influence on trade, 97; Kassites',
91; limitations of, 121; petroglyphs
of, 87; proto-, 72; religious aspects
of, 111–12; of Shang dynasty,
117–18, 119; speed of, 84; in steppe
societies, 86–89, 94; technological
advantage of, 91; technological
improvements for, 93; textual
references to, 90; transport of
crops, 98; troop transport by, 91;

Tutankhamun's, 86, 92–93;
two-team, 84, 131; warrior
superclass and, 97–98; in
Xiongnu, 141. *See also* transport,
horse; warfare; wheels
Chauvet (France), horse artwork of,
22–23
Chemurchek culture (Mongolia):
pastoralism of, 104; ritual sites, 104
Cheyenne people (Great Plains),
southern migration of, 191
Chichimec people (Mexico),
acquisition of horses, 184
Chile: domestic horses in, 196;
migrations from, 203; rock art of,
196
China, 116–20; agricultural lowlands
of, 119; Belt and Road Initiative,
222; civil war in, 140; climate
change in, 153–54; conflict with
steppe groups, 144–45; domestication in, 33; Great Wall, 144; horse
culture of, 112–14, 116–20, 137–39;
horse-raising difficulties, 119;
horse supply for, 119, 145–46;
resistance to riding, 138; steppe
culture in, 116–20; steppe raids in,
138; stirrup development in, 151,
152; trade with Himalayas, 147–48;
unification of, 138–39. *See also*
Shang dynasty
Chincoteague ponies, Spanish
ancestry of, 182
Chorrillo Grande, horse butchery at,
199
Cimmerians, conquest of Anatolia,
128
cities, steppe: archaeology of, 155,
156; capitals, 154–56

hypothesis for, 39–40, 43, 45, 50, 51, 62. *See also* human/horse relationships

Dominican Republic, introduction of horses to, 181

donkeys (*Equus asinus*), 64; adaptation of cattle carts to, 78; agricultural use, 66; domestication of, 64, 73, 121; iconography of, 69; riding of, 69, 73; of Spanish colonial world, 194; "territorial" system of, 12; transport use, 64–66, 78

Drews, Robert: *Coming of the Greeks*, 94

driving teams, soothing effect of, 79

DSK culture. *See* Deer Stone-Khirigsuur (DSK) culture

Durrington Walls (Great Britain), 171; chariots of, 89

durvuljin script, Mongol use of, 156

Eastern Seaboard, North American: Indigenous horse traders in, 192; introduction of horses to, 183, 186–87

Ede-Ile (Yorubaland), archaeological horse remains from, 164, *plate 16*

Egypt: Arabian migrants to, 162; chariots of, 92–93; donkey transport in, 64–65; horseback riding in, 122; Hyksos invasion of, 92; introduction of horses into, 160. *See also* Rameses II (the Great)

English Channel, transport of horses across, 171

Enriquillo (Taino leader), use of horses, 183

environmental change: anthropogenic, 37; effect on wild horses, 25–27. *See also* climate cooling; climate warming; deglaciated areas

Eohippus. See dawn horse

equids: in Andean civilizations, 194–95; of Dmanisi, 18; genomic analysis of, 70; of the Levant, 70; of Mesopotamia, 70; riding of, 68–69, 71*fig.*, 81, 122; skeletal differences among, 69; of South America, 15, 25, 194–95; transport using, 64–74; of western Asia, 66. *See also* asses, wild; donkeys

Equus (genus): divisions of, 14; genomic research on, 13–14; group social structure of, 12; hoof of, 11–12; of Last Glacial Maximum period, 23–25; male competition in, 12; Pleistocene radiation of, 15, 23; radiation across North America, 14–15; slender-legged, 14; stamina of, 12; westward spread of, 17. *See also* domestication of horses; horses

Equus caballus: control of, 84; evidence at Botai for, 49; skeletal recognition of, 35. *See also* horses, caballine

Equus ferus, population collapse of, 26

Equus ferus ferus, 76

Equus hemionus, subspecies of, 66. *See also* hemiones

Equus hydruntinus, 66

Equus lenensis, 27, 116

Equus neogeus, South American: hunting of, 25

mural, *plate 14*; Xiongnu conquest of, 140

Gelintang (Tibet), horse burial of, 146–47

genetic diversity: in domestication, 30; loss of, 220. *See also* DNA

Genghis Khan, 155

genome, equid: of Turkey, 91

genome, horse, 220; of Arabian caliphates, 162; of East Asia, 127; of Eurasia, 75

genome, human: of Inner Asia, 126; Yamnaya, 61

Georgia (republic), fossil record of, 18

ger tereg (yurt carts), 112

Ghana, horses of, 163

Giant Horse Gallery (Queensland), 211*fig.*, 212

globalization: of grasslands, 97; horses' role in, 98, 148; of steppe empires, 157; of steppe horse cultures, 134

goats, domestic, 32, 33*fig.*; genomic sequencing of, 32

Gobi Desert: burial practices of, 116; horse culture of, 116; during Pleistocene, 103

Godin Tepe (Iraq), hemiones of, 70

God Mol (Xiongnu tomb complex), Roman artifacts of, 141

Gran Chaco (grassland, South America), domestic horses in, 197

grasses: cell walls of, 11; damage to teeth, 8; digestion of, 9–11, 15; energy from, 9–11; environment for, 7–8

grasslands: African, 165; evolution of, 7–8; exposure to predators in,

8; globalization of, 97; human/horse relationships of, 74; insolation in, 153; proto-horses of, 9; Siberian, 23–24; temperature swings in, 8; water sources, 11. *See also* Great Plains, American; steppes

grasslands, South American: domestic horses in, 197–98

Great Britain: colonization of Australia, 207–10; Indo-Pacific expansion of, 207–8; passenger railway in, 216

Great Lakes: horse trade in, 192; introduction of horses to, 187

Great Plains, American: colonial pressures in, 191; Coronado's expedition to, 184; fencing of, 217; horse equipment of, 188–89; Horse Nations Indian Relay, 222; Indigenous American migrations to, 191; railways in, 217; smallpox epidemics in, 192; spread of horses to, 186. *See also* horse cultures, Indigenous North American; West, American

Greco-Bactrian tapestry pants, artist's reconstruction of, *plate 11*

Greece, ancient: bareback riders of, 149; effect of equestrian cultures on, 131; links to interior, 136; spread of chariots to, 94; threats from steppes, 128, 135

Greenland, Viking horses in, 178

guanaco (camelids, South America): mounted hunting of, 198–99, 204; trade in hides, 203

Guinnard, Auguste, 199–200

horse control: of Great Plains horses, 188; innovation in, xvi, 83–84; Persian innovations in, 135. *See also* bits; bridles; stirrups; transport, horse

horse cultures: Aboriginal, 210–12; of China, 112–14, 116–20, 137–39; economic aspects of, xiv; Eurasian, 112, 131; of Gobi Desert, 116; Indigenous scholars of, 223–24; of Koguryo state, 172; Māori, 212; of Mongolia, 107–16; in mythology, 121; Native Hawaiian, 212–13; of Oyo Empire, 164; Pazyryk, 132–34; Petrovka, 87–77; rapid relocation capabilities, 128; in religion, 121; Saka, 129, 130, 132; Sudanese, 164; of United States, 220–21; Viking, xvi, 175–79; in Western scientific literature, 223; of Xiongnu Empire, 142. *See also* Deer Stone-Khirigsuur (DSK) culture; human/horse relationships; Karasuk culture; Sintashta culture

horse cultures, Indigenous North American, 184–93; armor in, 189; buffalo hunting, 190–91; castration in, 188; colonial shocks on, 192; Comanche, 186; conflict in, 192; decoration of horses, 190; equipment, 188–89; exchange networks of, 192; horse lineages, 188, 222; horse meat in, 188; medicine and care in, 187; military successes of, 192; persistence under colonialism, 217–18; Plains horse equipment, 188–89; raw

materials in, 187–88; reproduction in, 188; rituals of, 190; seasonal migration in, 187; Shoshonean, 186; social systems of, 189; sovereignty in, 192–93; travois use, 189; wedding dowries in, 189

horse cultures, Indigenous South American, 196; animal wealth of, 203, 204; castration in, 199; gender dynamics of, 203; horseback hunting, 198–99, 204; horseback riding in, 195–98; horses as foodstuff in, 199–200; horse tack and goods, 201, 202*fig.*, 203; knowledge of horse behavior, 201; of the Pampas, 198–200; of Patagonia, 198–200; raiding in, 203; of Southern Cone, 199, 201, 203; spirituality of, 200; trade with Europe, 203; use of horse materials, 201; veterinary practices of, 200. *See also* Indigenous Americans (South America)

horse cultures, steppe, 98, 103, 120–21, 127–29; archeology of, 131; botanical remains of, 131; chariots and, 86–69, 94; confederacies of, 149; consolidation into empires, 143, 149; diversity of, 131–32; effect on settled societies, 131; expansion of, 126; farming in, 131, 132; globalization of, 134; innovations in, 142, 149–53; of Khangai mountains, xiii–xiv; in Korean Peninsula, 172; links to Japan, 173; mythologizing of, 131; persistence of stereotypes, 131; reshaping of

ski's horse, 15; technological data for, 90; of Tibet, 146–47; transformative force of, 77, 97–98; trans-Saharan, 162–64, 171; in 21st century, 220–24, *plate 20*; veterinary care for, 34, 42, 115–16, 134, 200; Viking, xvi, 175–79; versus wild, 29; during World War I, 217. See also *Equus*; horse cultures; human/horse relationships; transport, horse

horses, wild: of Australia, 209; body size of, 37; versus domestic, 29; effect of environmental change on, 25–27; friendships among, 13; of Great Plains, 191; in Hispaniola, 182; in human lifeways, 28; human predators, 78; hybrid with DOM2 domestic horse, 76; of Iberia, 15; mass harvesting of, 20, 23; population peak of, 15; slaughter in American West, 217; slaughter under industrialization, 218; of South America, 201, 203; westward spread of, 17. *See also* hunting of horses

horse stone monuments (Morin Mount valley), hoofprints of, xv

horse supply: Alexander the Great's, 136; for China, 119, 145–46; Himalayan, 146–47; in steppes, 119, 121, 143, 144; of Xiongnu, 140, 144, 146; of Yuezhi, 145; Zhang Qian's search for, 145

horse trade: along Tea Horse Road, 147–48, 205; Byzantine, 179; global, 179; Indigenous American, 192; transpacific, 207; trans-Saharan, 162. *See also* trade

human/horse relationships: archaeological records of, 18–19, 106–7; civilization change through, 98; during climate warming, 24–26; communication in, 13, 42; companionship in, 222; domestication in, 34–35; farming innovations in, 59; of grasslands, 74; Indigenous North Americans', 183–91; introduction of riding in, 114; predator and prey, xvi, 21; removal of biases from, 223–24; stages of, xvi; in 21st century, xvi, 221–24. *See also* domestication of horses; herders; horse cultures; hunting; riding

Huns, invasions of eastern Europe, 174

hunters, Paleolithic: artwork of, 22; permanent villages of, 32; strategies of, 20–21, 23, 28

hunting, horseback: of buffalo, 190–91; in South America, 198–99

hunting of horses, 21*fig.*, 78; archaeological record of, 18–21, 35; in Black Sea region, 76; at Botai, 48; burial assemblages for, 41; choice of location for, 53; in deglaciated areas, 24–26; demographics of, 20–21; in Eurasia, 36, 44; family group, 47*fig.*; harems, 21; mortality profiles for, 47*fig.*; Neanderthals', 19; Paleoindian, 24–25; Paleolithic traditions, 20–21, 48, 53; seasonal, 20; in South America, 25; strategies for, 20–21, 23, 28. *See also* horses, wild

Kassites (Zagros Mountains), charioteers of, 91

Kazakhstan: archaeological horses from, 127; domestic horses in, 89; horse herders of, 40, 85; population flow into, 126; spread of pastoralists into, 86

Kelly, Ned, 208

Khaganate, Turkic: capital cities of, 155; control of Eurasian steppes, 152–53; effect of climate cooling on, 153; runic script of, 156

Khangai mountain range (Mongolia): Afanasievo culture of, 64; horse culture of, xiii–xiv; Shatar Chuluu culture of, 61

Kharkhorum (Mongol capital): archaeozoology of, 156; roads to, 157

khirigsuurs (stone burial mounds, Mongolia), 115; horse sacrifices of, 109; layout of, 111–12; men and women in, 115

Khitan people, Chinese script use, 156

Khoikhoi peoples (South Africa), possession of horses, 206

Khovd (Mongolia), frame saddle of, 151–52

Khukh Nuur (Mongolia), iron stirrup of, 151

khulan, Mongolian, 15

kiang, Tibetan, 15

Kiowa people, of Great Plains, 191

Kivik (Sweden): archaeology of, 171; depiction of chariots at, 89

Kofun culture (Japan), horses in, 173

Koguryo state, horse culture of, 172

Korea: horses and horse equipment in, 172; long-haul voyaging technology, 172; stirrup development in, 151, 152

koumiss (fermented milk), 34, 114

K-Pg Extinction event. *See* Extinction, Cretacious-Paleogene

Krivyanskiy (Russia), domestication of horses at, 76–77

Kura-Araxes culture, cattle remains from, 60

kurgans (burial tumuli), 39, 130; wheeled wagons in, 60

lactose intolerance, human, 86

Lake Champlain, "horseboat" ferries of, 216

Lakota nation: geographic dominance of, 192; 21st century horse conservation, 222

land bridges: Alaska-Siberia, 23; dispersals across, 15, 17, 23

langur monkeys, bachelor groups, 12–13

L'anse aux Meadows (Newfoundland), Viking settlement at, 178

Last Glacial Maximum (LGM) period: extinctions following, 27; Mongolia following, 103–4

Late Helladic III period, horse burials of, 93

Lendbreen (Norway), horse snowshoe from, *plate 18*

Lesotho, horse rock imagery of, 206

the Levant, equids of, 70

Levine, Marsha, 40

lip rings, 83; for equids, 72–73

Littauer, Mary, 41

Little Big Horn (Montana), US cavalry defeat at, 217

social transformations, 203; spiritual traditions of, 200

Pawnee people: horse culture of, 189; horse traditions of, 186

Pazyryk culture (Altai Mountains): branding of horses in, 133–34; global integration of, 133; horse burials of, 133–34; material culture of, 132–33; mounted combat in, 132; proto-saddles of, 134–35, *plate 9*. *See also* Altai Mountains

Peloponnese, horse and chariot spread to, 93

Percy, George, 186

perissodactyla: cecum of, 10; of Eocene, 6; evolution of, 8

permafrost: preservation of DNA in, 27; tombs below, 141

Persia: animalist art of, 126; *onager* of, 15

Persia, Achaemenid: horse technology of, 135–36

Peru, Spanish: domestic horses of, 195–96

petroglyphs: Aboriginal, 211–12, 211*fig*.; of Altai Mountains, 86–87; of Ayrmach-Too, *plate 13*; horseback riding on, 113; horse-drawn chariots in, 87; Indigenous American, 193*fig*.; pan-Eurasian, 117. *See also* rock art

Petrovka culture (Russian-Kazakh border): bridle remains of, 87–88; chariot burials in, 88

Philippines, introduction of horses to, 207

Pico dos Ginetes, 180

pigs, in sedentary lifestyles, 60

Pizarro, Francisco: conquest of Inka Empire, 194

plague, 180; harbored by steppe marmots, 154; Justinian, 154

Plata, southern: feral horses of, 198

Pleistocene era: extinctions of, 27; Gobi Desert during, 103; radiation of *Equus* during, 15, 23; reversal of decline following, 39; temperature fluctuations in, 14, 23

plowing, cattle used for, 59

police, mounted, 220

Pompeii, ceremonial chariot of, 136, *plate 10*

Ponce de León, introduction of horses, 183

Pony Express (American West), 216

Potosí (Bolivia), equid bones of, 195

proto-horses: anatomical adaptations of, 15; behavioral adaptations, 12–13, 15; digestive system of, 10; DOM2 lineage, 75–76; evolutionary lines of, 13; hypsodont, 10; ligament structure, 11; predator evasion, 11, 15; speed and movement of, 11–12, 15; survival adaptations of, 9; water sources for, 11. *See also* dawn horse; equids; horses, caballine

proto-primates: carnivore, 16; evolution in Africa, 7; of Paleogene radiation event, 6

Przewalski's horse, 15, 103; aggressive traits of, 51; of Altai Mountains, 107; artistic depictions of, 68; at Botai, 51; coat patterns of, 22; harem system of, 12; of Holocene China, 116; skeleton of, 35; split of domestic horses from, 15

Pueblo people: conflicts with Coronado, 184; Revolt (1680), 185; Spanish *visitas* to, 185

Puelche people (Southern cone), horse tack of, 201

Puerto Real (Haiti), horse artifacts from, 182, *plate 19*

Qin Shihuang, Emperor: mausoleum of, 139

Qin state (China): adoption of cavalry, 138–39; Great Wall of, 144; Xiongnu conflict, 140

Quanrong ("dog barbarians"), sacking of Zhou capital, 138

Queensland, Aboriginal horse culture of, 211

radiocarbon dating, 223; at Deri-yevka, 44; dietary factors affecting, 76–77; of horse burials, 110

railways: in American Great Plains, 216, 217; effect on Native Americans, 217; horses' coordination with, 216

rainforests: barriers to domestication of horses, 164–65; drying of, 14

Rameses II (the Great), Pharaoh, 160; defeating the Hittites, 122, *plate 7*; stable of, 93

Recht, L., 67

reliefs: of Assyrian cavalry, 125, 125*fig.*; of Borobodur, 206

reproduction. *See* breeding

rhea, South American: horseback hunting of, 199, 204, 222

riding: in Arabia, 160; for athletic performance, 122; bareback, 134,

149; in battle of Kadesh, *plate 7*; in combat, 122, 124–26, 125*fig.*; control in, 122–24; distribution of weight, 69, 81; "donkey seat," 69, 71*fig.*, 81; in DSK culture, 112–14, 123; economic power in, 149; effect in ancient world, 129–30; in Egypt, 122; ethos of wealth in, 115–16; in Eurasia, 142–43; at Hasanlu, 125–26; in herding, 127–28; Indigenous South Americans', 195–98; in Inner Asia, 121, 126–27, 131; leg grips, 124; messengers', 122, 124, *plate 7, plate 14*; in Mongolian steppes, 114–15, 120–21, 123–24, 126; overcoming aggression in, 78–79; in pairs, 124–25; petroglyphs of, 113; political power in, 149; selective breeding for, 123; skeletal evidence for, *plate 4*; social impact of, 129–30; speed in, 73; spread to China, 116; technological improvements in, 122–23; trade routes and, 130; transformational aspects of, xvi; trousers for, 113; westward spread of, 128, 129; Xiongnu, 142. *See also* bridles; cavalry; saddles; stirrups

Rig Veda hymns, domestic horse in, 95

Río de la Plata (South America), Spanish colonists at, 197

road systems: horse-based, 222; of Mongol Empire, 157, 158; Silk Road, 148, 222; Tea Horse Road, 147–48, 205

rock art: Asian, 86–87; of Chile, 196; of Lesotho, 206; North American,

transport, donkey, 64–66; advantages of, 65; in Egypt, 64–65; in Near East, 65–66, 77–78; ring-and-rein use, 84; technology of, 70–71, 73

transport, equid, 64–74; archaeozoological record of, 69–70, 73; battle cars, 71*fig.*, 74; braking, 72; control in, 72–73, 81, 83–84; first, 64–65; iconographic depictions of, 67–68; for long-distance communication, 73–74; ring-rein system of, 72–73; skeletons of, 72; socioeconomic changes accompanying, 64; on Standard of Ur, 67; straddle cars, 71–72, 71*fig.*; technology of, 70–73; zebra, 80

transport, horse, 34; across Panama, 182; in Australia, 208; control of horse herds, 85; damage to teeth in, 41–42, 43; on deer stone monuments, 110–16; desert, 160; of DSK culture, 110–16; global impact of, 214; in Great Mongol Empire, 157, 158; impact on social status, 97; Indigenous American, 189; in Karasuk culture, 105–6; limitations of, 121; in North Africa, 92, 161; origin of, 77–78; replacement by mechanization, 218–19; research methods for, 51–52; in Sintashta culture, 82–83; sledge, 59; spread of, 56–57*map*; using teams, 79, 121. *See also* chariots; riding

trans-Saharan regions: horse equipment of, 163–64; horse trade of, 162; trade in, 163–66; travel corridors of, 163–64. *See also* Sahara, northern

travel routes: horse-based, 130, 222; trans-Saharan, 162–64. *See also* road systems

treadwheels, horses working, 216

Troy, bridle cheekpieces from, 94

trypanosomiasis, 165

Tsagaan Asga (Mongolia), livestock of, 105

tsetse fly, 165

Tunning 1 (Russian Tuva), DSK/ Central Asian influences in, 126–27

Turkey, equid genome of, 91

Turkmenistan, bridle artifacts from, 96

Tutankhamun, Pharaoh, 160; chariots of, 86, 92–93

Tyagunova Mogila Cemetery (Ukraine), two-wheeled cart from, 82*fig.*

Uai Bobo (Timor), horse archaeology of, 209

Ulaanbaatar (Mongolia): horse traffic light, *plate 20*; Naadam festival, 220

Umm el-Marra (Syria): equid skeletons of, 70, 72; equid teeth from, 70, *plate 6*

Únětice culture (Poland), caballine horses at, 8

United States, horse culture of, 221

Ural region (Russia), horse skeleton assemblages of, 38

Urd Ulaan Uneet (western Mongolia), frame saddle of, 152, *plate 15*

Uttar Pradesh (India), carts of, 95

Founded in 1893,
UNIVERSITY OF CALIFORNIA PRESS
publishes bold, progressive books and journals
on topics in the arts, humanities, social sciences,
and natural sciences—with a focus on social
justice issues—that inspire thought and action
among readers worldwide.

The UC PRESS FOUNDATION
raises funds to uphold the press's vital role
as an independent, nonprofit publisher, and
receives philanthropic support from a wide
range of individuals and institutions—and from
committed readers like you. To learn more, visit
ucpress.edu/supportus.